OTHER PUBLICATIONS BY THE ASSOCIATION

Makedonika: Essays by Eugene N. Borza
Edited by Carol G. Thomas

The Coming of The Greeks
James T. Hooker

Directory of Ancient Historians in the United States, 2nd ed.
Compiled by Konrad Kinzl

The Legacy of Ernst Badian
Carol G. Thomas, T. Corey Brennan, Stanley M. Burstein, Eugene N. Borza,
Jerzy Linderski

Continued publication of the series is made possible through the efforts of the AAH publications committee, Peter Bedford of Union College; Eugene Borza of the Pennsylvania State University; Serena Connolly of Rutgers, The State University of New Jersey, president; Danielle Kellogg of Brooklyn College—City University of New York; Hans-Friedrich Mueller of Union College, chair; Isabelle Pafford of San Jose State University; Jonathan Scott Perry of the University of South Florida—Sarasota-Manatee; David Ratzan of the Institute for the Study of the Ancient World, New York University; and Jennifer T. Roberts of City College—City University of New York. Readers with questions about the series or topic suggestions for future volumes or manuscript questions should contact the current president of the Association of Ancient Historians.

NEW DIRECTIONS IN THE STUDY
OF ANCIENT GEOGRAPHY

Publications of the Association of Ancient Historians 12

PUBLICATIONS OF THE
ASSOCIATION OF ANCIENT HISTORIANS

The purpose of the monograph series is to survey the state of the current scholarship in various areas of ancient history.

New Directions in the Study of Ancient Geography

Publications of the Association of Ancient Historians 12

EDITED BY DUANE W. ROLLER

Association of Ancient Historians

Library of Congress Cataloging-in-Publication Data

Names: Roller, Duane W., editor.
Title: New directions in the study of ancient geography / edited by Duane W. Roller.
Other titles: Publications of the Association of Ancient Historians.
Description: Sarasota, Florida : Association of Ancient Historians, [2020] | Series: Publications of the Association of Ancient Historians series | Includes bibliographical references and index.
Summary: "A collection of essays on current studies in ancient geography, extending over an area from ancient Mesopotamia and the prehistoric New World to the Roman Empire. Essays include examinations of ancient cosmology, ancient navigation, and literary interpretations of geography"—Provided by publisher.
Identifiers: LCCN 2019050766 | ISBN 9781734003116 (cloth) | ISBN 9781734003109 (paperback)
Subjects: LCSH: Geography, Ancient.
Classification: LCC G84.N48 2020 | DDC 910—dc23
LC record available at https://lccn.loc.gov/2019050766

Eisenbrauns is an imprint of The Pennsylvania State University Press.

The Pennsylvania State University Press is a member of the Association of University Presses.

It is the policy of The Pennsylvania State University Press to use acid-free paper. Publications on uncoated stock satisfy the minimum requirements of American National Standard for Information Sciences—Permanence of Paper for Printed Library Material, ANSI Z39.48–1992.

CONTENTS

ABBREVIATIONS

Primary Texts

Apollonios, *Argon.*	*Argonautica*
Diodoros of Sicily, *Hist.*	*Histories*
Diogenes Laërtios, *Lives*	*Lives and Opinions of Eminent Philosophers*
Herodotus, *Hist.*	*Histories*
Hesiod, *Theog.*	*Theogony*
Homer, *Il.*	*Iliad*
Homer, *Od.*	*Odyssey*
Josephus, *J.W.*	*Jewish Wars*
Manilius, *Astron.*	*Astronomica*
Pausanias, *Descr.*	*Description of Greece*
Pliny, *Nat. Hist.*	*Natural History*
Pomponius Mela, *Chor.*	*Chorography*
Strabo, *Geogr.*	*Geography*
Thucydides, *Hist.*	*History of the Peloponnesian War*
Vergil, *Aen.*	*Aeneid*

Secondary Sources

AC	*Antiquité Classique*
ACD	*Acta classica Universitatis Scientiarum Debreceniensis*
AJPh	*American Journal of Philology*
AncW	*Ancient World*
AOF	*Archiv für Orientforschung*
ArchPhilos	*Archives de philosophie*
BICS	*Bulletin of the Institute of Classical Studies*
BNJ	*Brill's New Jacoby*
CIL	*Corpus inscriptionem latinarum*
CJ	*Classical Journal*

CP	*Classical Philology*
CQ	*Classical Quarterly*
CR	*Classical Review*
CSCA	*California Studies in Classical Antiquity*
CW	*Classical World*
DK[6]	Hermann Diels and Walther Kranz, *Die Fragmente der Vorsokratiker*. 6th edition. Berlin: Weidmann, 1951.
EANS	*Encyclopedia of Ancient Natural Scientists: The Greek Tradition and Its Many Heirs*. Edited by Paul T. Keyser and Georgia L. Irby-Massie. London: Routledge, 2008.
FGrHist	*Fragmente der Griechischen Historiker*
GLM	*Geographi Latini Minores*. Edited by Alexander L. Riese. Heilbrun 1878; repr. Hildesheim: Olms, 1964.
GRBS	*Greek, Roman, and Byzantine Studies*
GRL	*Geschichte der Römischen Literatur*. 4 vols. Edited by M. Schanz and C. Hosius. Munich: Beck, 1896–1934.
HR	*History of Religions*
HSCP	*Harvard Studies in Classical Philology*
ILS	*Inscriptiones latinae selectae*
JAOS	*Journal of the American Oriental Society*
JEA	*Journal of Egyptian Archaeology*
JHI	*Journal of the History of Ideas*
JHS	*Journal of Hellenic Studies*
JIES	*Journal of Indo-European Studies*
JNES	*Journal of Near Eastern Studies*
JNG	*Jahrbuch für Numismatik und Geldgeschichte*
JRGZ	*Jahrbuch des Römisch-Germanischen Zentralmuseums*
JRS	*Journal of Roman Studies*
JThS	*Journal of Theological Studies*
LÄ	*Lexikon der Ägyptologie*. Edited by Wolfgang Helck and Eberhard Otto. 8 vols. Wiesbaden: Harrassowitz, 1975–1990.
MD	*Materiali e discussioni per l'analisi dei testi classici*
MDAI(R)	*Mitteilungen des Deutschen Archäologisches Institut, Römische Abteilung*
MH	*Museum Helveticum*
NECJ	*New England Classical Journal*
PCA	*Proceedings of the Classical Association*
PCPhS	*Proceedings of the Cambridge Philological Society*
PVS	*Proceedings of the Virgil Society*
RE	*Real-Encyclopädie der classichen Altertumswissenschaft* (Pauly-Wissowa)

RFIC	*Rivista di filologia e di istruzione classica*
RIB	*The Roman Inscriptions of Britain.* Edited by R. G. Collingwood and R. P. Wright. Stroud: Sutton, 1995.
SHA	Scriptores historiae Augustae
SIFC	*Studi italiani di filologia classica*
SyllClass	*Syllecta Classica*
TAPhA	*Transactions of the American Philological Association*
TAPhS	*Transactions of the American Philosophical Society*
TEGP	Daniel W. Graham, *The Texts of Early Greek Philosophy: The Complete Fragments and Selected Testimonies of the Major Presocratics.* Cambridge: Cambridge University Press, 2010.
YCS	*Yale Classical Studies*
ZÄS	*Zeitschrift für ägyptische Sprache und Altertumskunde*
ZPE	*Zeitschrift für Papyrologie und Epigraphik*

Introduction

Duane W. Roller

GEOGRAPHY HAS BEEN AN ESSENTIAL PART of the human condition ever since the first primitive people wandered from one place to another and encountered new mountains and rivers that changed their visual horizon or felt the astonishment of reaching a previously unknown seacoast. These early experiences could hardly be called geography in any intellectual sense, but they created an awareness of the vast expanse and varied nature of the surface of the earth.

As society became more complex, travellers went farther from home and began, perhaps almost unconsciously at first, to gather ethnographic and topographical information that could be useful. Since the earliest travellers tended to be merchants or military personnel, details about the people they met and the landforms and seas they saw were valuable data. In time, such material began to coalesce into the rudiments of geographical fact and theory. In the Greek world, sailors took the lead and were the first to go the farthest: even in the enclosed Mediterranean and Black Sea system there were many harbors to explore and many peoples to encounter. They realized that the earth was curved—some even dared to suggest it was a sphere—and, moreover, that they could use the positions of celestial bodies to determine their location. Eventually Greeks moved beyond the data collection stage to form a simple theory of geography. As merchants, soldiers, and sailors reported back to their home locations about what they had seen, mathematicians and philosophers began to create theoretical geography, based on concerns about the size and shape of the earth, its relationship to the celestial sphere, and the relevance of the changing climate and length of day to one's movements around the earth. Thus the development of ancient geography as an intellectual discipline proceeded along two paths: collection of information and theorization about its meaning.

The textual record of Greco-Roman geographical information originates with the beginning of Greek literature. Nearly every ancient author has some geographical material, even long before there was any formal discipline of geography or scholarly writing on the topic. Geographical notices in Greek literature began with the sailing instructions given by Kalypso to Odysseus and include Hesiod's knowledge of the coasts of the Black Sea and the astonishingly extensive topographical data in the plays of Aeschylus. All of these details reflected the extent and expansion of contemporary knowledge: Homer's world was limited to the span from Anatolia to Italy and south to Libya, Hesiod learned about the rivers of central Asia, and Aeschylus could report on the western Mediterranean.

Natural philosophy, geology, astronomy, topography, and ethnography were the component parts of scholarly geography. By the fourth century BC, there was a movement toward collating these disparate parts into a new discipline. Plato, Eudoxos of Knidos, Ephoros, and Aristotle all contributed in their own way to this effort, but there was no treatise solely devoted to the topic of geography until that of Eratosthenes of Kyrene, written in the second half of the third century BC. Eratosthenes used his mathematical skills to determine the size of the earth and then used geographical information to locate hundreds of places on it. Moreover, he invented the word *geographia* to describe his endeavors.

After Eratosthenes, geography took its place among the scholarly disciplines and, especially in later Hellenistic times, became a frequent subject of treatises. But, as is so often the case with scholarly writings from antiquity, the loss of such works has been tremendous: nearly 250 Greek and Roman geographers are known by name, but only the writings of four have survived. Even Eratosthenes's seminal treatise was lost by the second century AD and only about 150 fragments of it remain.

The earliest extant geography is the *Geography* of Strabo of Amaseia, completed in the early first century AD. Its seventeen books (one of the longest surviving works of Greek literature) cover the entire history of geographical thought previous to his day; without it, virtually nothing would be known about the topic. A generation later, Pomponius Mela wrote his *Chorography* (completed in the 40s AD). Important as the only extant independent geography in Latin is, its lack of detail (only three books) can be frustrating. Completed in the 70s AD was the *Natural History* of Pliny the Elder, whose geographical books (2–6) are the most complete discussion of the topic in Latin. Around the middle of the second century AD, Ptolemy of Alexandria completed his *Geographical Guide*, more a work on mathematics and mapmaking than pure geography, but nonetheless exceedingly valuable.

Despite the great loss of geographical texts from the Greco-Roman world, the study of ancient geography became of immense importance in the Renaissance, as explorers, such as Columbus, used Strabo, Pliny, and Ptolemy to plot and interpret their far-ranging voyages. Works on the history of ancient geography began to appear in the nineteenth century: in English, the first important study was E. H. Bunbury's *History of Ancient Geography* (1883). One can also include J. Oliver Thomson's *History of Ancient Geography* (1948) and the present editor's *Ancient Geography* (2015).[1] Yet the study of ancient geography is a rapidly evolving discipline: it is only in the last two decades that reliable critical texts and commentaries have appeared of the four extant authors (some of this work is still in progress), and sophisticated mapmaking techniques have been brought into play to interpret the topographical data.

This volume contains five essays, arranged chronologically, that cover the period from the ancient Near East into Roman imperial times and reflect current thoughts and trends in geographical scholarship. They demonstrate the diversity yet continuity implicit in the study of ancient geography, where there is a trajectory from the origins of civilization to the Roman period. Paul Keyser's "Kozy Kosmos of Early Cosmology" starts at the beginning with consideration of the universe itself and how seminal Near Eastern views affected Greek scholars, who held attitudes toward the cosmos that are still relevant today. Hundreds of years later, as seamen, merchants, and traders built on the collective wisdom gained by innumerable generations of developing geographical knowledge, geography became a useful technical tool of imperial ideology. In the early third century BC, when Ptolemaic Egypt became the most powerful state in the Hellenistic world, its officials needed accurate knowledge of coastlines, ports, and harbors that would be visited by their naval and mercantile vessels. The editor of this volume has offered a text and commentary on Timosthenes of Rhodes, who was chief of naval staff for Ptolemy II and produced a navigational guide for Ptolemaic seamen, which survives only in fragments. Although little remains of Timosthenes's treatise, it is representative of what must have been a common geographical genre in antiquity, as trade and military adventurism became ever more common.

An essential by-product of geography was cartography. It is also in itself an implementation of geographical scholarship. The first maps in the Greek world seem to date from the sixth century BC, and the genre of mapmaking took many forms, literary as well as physical. Yet few actual maps survive from antiquity. There is a symbiotic relationship between physical maps and literary descriptions of them, and it is upon the latter that modern scholars must largely rely for much of their understanding of the former. Georgia L. Irby has provided an analysis of one unusual but familiar map from literature, the Shield of Aeneas in the eighth book of Vergil's *Aeneid*.

As noted, extant geographical treatises from the Greco-Roman world are few. The earliest to survive in Latin is the *Chorography* of Pomponius Mela,

---

1. E. H. Bunbury, *History of Ancient Geography* (London: Murray, 1883); J. Oliver Thomson, *History of Ancient Geography* (Cambridge: Cambridge University Press, 1948); Duane W. Roller, *Ancient Geography: The Discovery of the World in Ancient Greece and Rome* (London: Tauris, 2015).

from the middle of the first century AD. In her second essay in this volume, Georgia L. Irby has produced the first modern and detailed study of this author and his work, bringing into contemporary understanding the only extant and independent geographical work in Latin.

From its earliest days, geography has served national interests, and even in modern times it continues to be part of state ideology. The final essay in this volume, by Molly Ayn Jones-Lewis, describes how Tacitus's *Germania*, dated to the end of the first century AD, is a work heavily reliant on environmental determinism, an issue still highly relevant today.

These essays cover many generations of ancient geographical scholarship and demonstrate the great diversity of the discipline, both in antiquity and in modern times. Extending from the era of primitive cosmology to the ideology of the Roman Empire, they offer strong proof of the extensive new directions taken by modern scholars of ancient geography.

The editor would like to thank those who assisted in the preparation of the volume and the other contributing authors for their hard work. Particular thanks are due to the Association of Ancient Historians for commissioning this volume, and especially its successive editors of publications, Serena Connolly and Hans-Friedrich O. Mueller. The authors have also provided acknowledgments in their texts.

CHAPTER I

The Kozy Kosmos of Early Cosmology

Paul T. Keyser

ARISTOTLE EXPLOITED PRIOR SPECULATIONS about the *kosmos*—claims about the elements, the shape of the earth, the heavenly motions, and the causes of things—in order to support his own claims, which he considered the *telos* of those earlier works. But a mass of evidence exists that those earlier thinkers were speculating within, or on the margins of, a radically distinct framework, which has left traces throughout early Greek thought and even within Aristotle's works. Rethinking that evidence reveals features of a model that had no tendency to evolve in the direction of Aristotle's model, and that I designate as the "Kozy Kosmos." In this model, the flat earth is bounded at its rim and beneath by watery chaos, with a sacred mountain, pillar, or tree at its center, and overlain by a rotating sky, whose changes are causally connected with changes on earth. This mytho-historical model is anthropocentric, constructed of contiguous and cohering parts, divinely ordered, and threatened by chaos.

To better reimagine this model, I exploit evidence of early cosmogonies and cosmologies from Egypt, Israel, Mesopotamia, India, China, and even the Americas. By confronting the reliable evidence for early Greek cosmology with the reimagined mytho-historical model, I show how radically transgressive and yet conservative the development of early Greek cosmology was. This perspective provides a clearer focus for many claims attributed to the writers Akousilaos, Alkman, Anaxagoras, Anaximander, Anaximenes, Antiphon, Aristeas, Demokritos, Diogenes of Apollonia, Empedokles, Herakleitos, Melissos, Parmenides, Pherekudes of Suros, Thales, Xenophanes, and Zeno of Elea. Furthermore, key concepts from this "Kozy Kosmos" model persist within the models of Plato and Aristotle. In a related sense, the models of Plato and Aristotle are also "cozy," in that they are stable, closed, and of human scale.[1]

1. Alan C. Bowen and Michèle Lowrie read and provided valuable advice on an early version of §III, "Rounding the Edges, Raising the Sky" (paper given in San Francisco at the 2004 Annual Meeting of the American Philological Association, now known as the Society for Classical Studies). Duane W. Roller encouraged that early work and its current expansion. For all of that support, I am very grateful.

I. The Mytho-historical "Cradle Cosmology": An Early and Discarded Image

The modern synthesis depicts a *kosmos* of stunning extent in space and time, vastly exceeding human grasp, and we find ourselves lost among its infinities. As Blaise Pascal (1623–1662) mused, "Le silence éternel de ces espaces infinis m'effraie."[2] This world-picture has evolved through debate and discovery since the publication by Nicolaus Copernicus (1473–1543) of the *De revolutionibus orbium coelestium* (1543). The vastness that terrified Pascal has grown ever more vast through the subsequent works of Isaac Newton (1643–1727 [new style dates]), Pierre-Simon Laplace (1749–1827), and Charles Darwin (1809–1882), plus the twentieth-century works of Albert Einstein (1879–1955), Arthur Eddington (1882–1944), and Edwin Hubble (1889–1953).

If we turn and gaze back at the evolution of the world-pictures of ancient cultures, we are amazed in a different way. The synthesis created in the Greek world by Plato and Aristotle became the received view, despite resistance by atomists, throughout cultures influenced by ancient Greco-Roman culture, until it was discarded in favor of the world-picture that began to emerge with Copernicus's work.[3] Because of the long life of, and extensive studies on, this culturally pervasive, geocentric, spherical-*kosmos*, four-or-five-element model, we are somewhat familiar with its assumptions and choices, but it is in fact quite peculiar and contingent. Why should there be four earthly elements that rise or fall, and one heavenly element that eternally rotates? Why should two planets bob about on their circles near the sun, and three others not do so? Why should there be five twinkling wandering stars, and two great luminaries, plus thousands of "fixed" stars? Why is there a swath of stars like dust in a band across the sky? Why do all those rotations happen at all? Indeed, some of these puzzles were raised in Greco-Roman antiquity.

But we can go further back, and further afield, and examine alternative models that were created in ancient Greek culture or in the ancient cultures of Egypt, Mesopotamia, India, China, or even the Americas. There also we are amazed to find a set of models very similar to one another. Moreover, these models illuminate the Greek models propounded before Plato, as well as some of the peculiar features within those early Greek models and even within the synthesis created in the Greek world by Plato and Aristotle. It is beyond the scope of this

---

2. Léon Brunschvicg, ed., *Blaise Pascal: Pensées* (Paris: Vrin, 1904), #206 (p. 127) = Michel Le Guern, *Pascal: Oeuvres complètes* (Paris: Gallimard, 2000), #187 (p. 615); for a PDF image of this pensée in the two MS copies, see: http://www.penseesdepascal.fr/C1-C2/C1p101-C2p129 -Transition7.pdf.

3. C. S. Lewis, *The Discarded Image: An Introduction to Medieval and Renaissance Literature* (Cambridge: Cambridge University Press, 1964).

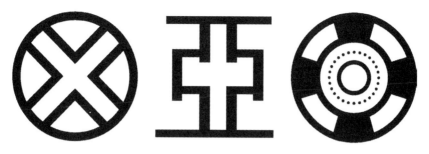

Egyptian determinative for city   Chinese 'ya'   Mayan diagram of Earth

FIGURE I.I. The Quincunx (Flat) Earth, courtesy Paul A. Whyman.

paper to explain the parallels, although horizontal transfer ("borrowing") seems highly unlikely, and instead they are more likely due to either an ancient common inheritance or something like convergent evolution—that is, similar human responses to similar human experiences.[4]

These primordial world-pictures are relatively well documented in early hieroglyphic texts from Egypt, and likewise in early cuneiform texts from Mesopotamia. The early Sanskrit texts were composed in centuries contemporaneous with Greco-Roman culture but reproduce material that scholars agree derives from the Vedic era of 1500–1000 BC. Similarly, the earliest Chinese texts that survive are contemporaneous with Hellenistic Greek and Greco-Roman culture, but they reproduce material that scholars trace back to the Shang dynasty (roughly 1600–1100 BC). Finally, the partly deciphered monumental texts of the Classic Maya period are dated to ca. 200–900 AD, and the few surviving codices are dated to ca. 1500–1700 AD, but scholars agree that they both reflect much older material. Making use of all these materials requires no hypothesis about (in)dependence—the only assumptions are that comparisons can validly be made and that those comparisons might be revelatory or illuminating. Let us consider those materials, in geographical order of origin, beginning with Egypt and proceeding eastward to the Maya.

Egypt

The Pyramid texts and later hieroglyphic texts depict a flat world, oriented south to north along the course of the Nile, surrounded by dangerous foreigners, and

4. Indeed some of the parallels seem to derive from human neurophysiology: the illusions that: (1) the moon and sun are larger near rising and setting, (2) the sky is a flattened dome, (3) the sky is spinning, and (4) crepuscular rays diverge.

divinely ordered through the pharaoh.[5] Below the land, or in some unspecified "other place," was the *dw3t* ("Duat"), where the sun went at night, a cosmic region of darkness, stars, and water, with its entrance in the northwest and its exit in the southeast.[6] Above the land, the sky provided fresh water and was a shining firmament (*bỉ3*), possibly made of the heavenly metal iron (*bỉ3t*),[7] and probably flat, as seen in the determinative for "sky" (*pt*): ⌐.[8] The world had been created out of dark and formless primordial water, *nwn* (masculine) and *nwnt* (feminine), when the creator god called the first mound (*q33*) into being by his word.[9] That originary mound generated Egypt itself, the center of the world, and was the model for every temple.[10] The symbolic shape of the earth in the *kosmos* was repeated in the hieroglyphic determinative for "city," a circle around a quincunx: ⊗.[11] There seems to be no world-tree or world-mountain (Egypt is a land of few and low mountains), but a distinctive trace of the world-pillar persists in the hieroglyph for the *ḏd*-pillar (𓊽), which depicts a cosmic sky-support in the Old Kingdom and is an emblem of stability.[12]

5. James P. Allen, "The Cosmology of the Pyramid Texts," in *Religion and Philosophy in Ancient Egypt*, ed. James P. Allen (New Haven: Yale Egyptological Seminar, 1989), 1–28; Jan Assmann, *Ägypten: Theologie und Frömmingkeit einer frühen Hochkultur* (Stuttgart: Kohlhammer, 1984), 84–90; Hellmut Brunner, "Zum Raumbegriff der Ägypter," *Studium Generale: Zeitschrift für die Einheit der Wissenschaften* 10 (1957): 616–617; Joanne Conman, "It's About Time: Ancient Egyptian Cosmology," *Studien zur Altägyptischen Kultur* 31 (2003): 36; Sakkie Cornelius, "Ancient Egypt and the Other," *Scriptura* 104 (2010): 322–40; Leonard H. Lesko, "Ancient Egyptian Cosmogonies and Cosmologies," in *Religion in Ancient Egypt: Gods, Myths, and Personal Practice*, ed. Byron E. Shafer (Ithaca: Cornell University Press, 1991), 117; Hans Schwabl, "Weltschöpfung," *RE*, 2nd ser. 9 (1962), §42, cols. 1499–502; Vincent Arieh Tobin, "Creation Myths," in *Oxford Encyclopedia of Ancient Egypt*, ed. D. B. Redford (Oxford: Oxford University Press, 2001), 2:469–72.

6. Allen, "Cosmology," 21–25; Conman, "It's About Time," 36–37, 42–43.

7. Allen, "Cosmology," 7–10; Dieter Kurth, "Nut," *LÄ* 4 (1982): 533–41; Lesko, "Ancient Egyptian," 117.

8. NI in the sign-list of Alan H. Gardiner, *Egyptian Grammar*, 3rd ed. (London: Clarendon, 1957), 485; cf. Dirk L. Couprie, *Heaven and Earth in Ancient Greek Cosmology* (Berlin: Springer, 2011), 3, 5. A domed heaven may be implied by later depictions that show a curved heaven, as argued by Couprie, 5–9.

9. Jan Assmann, "Schöpfung," *LÄ* 5 (1984), 678; Brunner, "Zum Raumbegriff," 615; Françoise Dunand and Christiane Zivie-Coche, *Dieux et Hommes en Égypte* (Paris: Colin, 1991; repr. 2001), 55–57, 60–61; trans. by David Lorton as *Gods and Men in Egypt: 3000 BCE to 395 CE* (Ithaca: Cornell, 2004), 45–47, 50–52; Maulana Karenga, *Maat, the Moral Ideal in Ancient Egypt: A Study in Classical African Ethics* (New York: Routledge, 2004), 191–192.

10. Brunner, "Zum Raumbegriff," 616; Cornelius, "Ancient Egypt," 324; Hans J. Klimkeit, "Spatial Orientation in Mythical Thinking as Exemplified in Ancient Egypt: Considerations Toward a Geography of Religions," *HR* 14 (1975): 270; Karl Martin, Urhügel," *LÄ* 6 (1986): 873–75; Siegfried Morenz, *Ägyptische Religion* (Stuttgart: Kohlhammer, 1960), 46–47, trans. by Ann E. Keep as *Egyptian Religion* (Ithaca: Cornell, 1973), 44–45.

11. Brunner, "Zum Raumbegriff," 618. The sign is O49 in the sign-list of Gardiner, *Egyptian Grammar*, 498.

12. Gardiner, *Egyptian Grammar*, 502, sign R11 in the sign-list; Robert T. R. Clark, *Myth and Symbol in Ancient Egypt* (London: Thames and Hudson, 1978), 236–237; H. Altenmüller, "Djed-Pfeiler," *LÄ* 1 (1975): 1100–1105; Raymond O. Faulkner, s.v. in *A Concise Dictionary of Middle*

The formless primordial waters lay at the edges of this divinely ordered world, perhaps to invade at any moment.[13] Egyptian texts persistently manifest such potent disdain for foreigners that Egyptologists describe it as xenophobia, which derived from an ideology of nationalism.[14] In order to prevent the influx of chaos, the Pharaoh had to rule and worship in accord with *mꜣ't* ("Ma'at"), meaning something like "truth, justice, and the Egyptian way," with connotations of balance, order, and harmony. That activity would preserve the order and prosperity of Egypt.[15] Scholars explicitly compare the Pharaoh acting in accord with Ma'at to the Chinese Emperor acting in accord with the Mandate of Heaven;[16] likewise, they compare the central position of Egypt in the *kosmos* to the central position of China in its *kosmos*.[17] Diodoros of Sicily (*Hist.* 1.30–31) and Josephus (*J.W.* 4.607–610) report the Egyptian perception of their homeland as a securely walled country: Josephus writes, "by land it is hard to enter and by sea harborless ... thus Egypt is everywhere walled" (κατά τε γῆν δυσέμβολος καὶ τὰ πρὸς θαλάσσης ἀλίμενος ... τετείχισται μὲν οὕτως ἡ Αἴγυπτος πάντοθεν).

Mesopotamia

The cuneiform texts record many variant creation stories, which tell parallel, but not always consistent, tales.[18] The creation began when there was nothing

Egyptian (Oxford: Oxford University Press, 1976), 325; Jennifer McKeown, "The Symbolism of the Djed-pillar in *The Tale of King Khufu and the Magicians*," *Trabajos de Egiptología / Papers on Ancient Egypt* 1 (2002): 57–58.

13. Hellmut Brunner, "Die Grenzen von Zeit und Raum bei den Ägyptern," *AOF* 17 (1955): 143–144; Erik Hornung, "Chaotische Bereiche in der geordneten Welt," *ZÄS* 81 (1956): 28–32; Brunner, "Zum Raumbegriff," 613–14; Morenz, *Ägyptische Religion*, 176–177; Morenz, *Egyptian Religion*, 168; Klimkeit, "Spatial Orientation," 279; Assmann, "Schöpfung," 685–86.

14. Antonio Loprieno, *Topos und Mimesis: Zum Ausländer in der ägyptischen Literatur* (Wiesbaden: Harrassowitz, 1988), 22–40; Toby A. H. Wilkinson, "What a King Is This: Narmer and the Concept of the Ruler," *JEA* 86 (2000): 28–29.

15. Jan Assmann, *Der König als Sonnenpriester: Ein kosmographischer Begleittext zur kultischen Sonnenhymnik in thebanischen Tempeln und Gräbern* (Glückstadt: Augustin, 1970), 62–65; Assmann, *Ägypten*, 84–90 / 68–73; Jan Assmann, *Maât: L'Égypte pharaonique et l'idée de justice sociale* (Paris: Julliard, 1989), 127–28; Cornelius, "Ancient Egypt," 326, 330; Phillipe Derchain, "Le rôle du roi d'Égypte dans le maintien de l'ordre cosmique," in *Le pouvoir et le sacré*, ed. Luc de Heusch (Brussels: Université libre de Bruxelles, 1962), 68–72; Dunand and Zivie-Coche, *Dieux et Hommes en Égypte*, 61–62; trans. *Gods and Men in Egypt*, 52–53; Henri Frankfort, *Kingship and the Gods*, 2nd ed. (Chicago: University of Chicago Press, 1978), 51–60, cf. p. vii; Karenga, *Maat*, 30–34; Klimkeit, "Spatial Orientation," 271; Morenz, *Ägyptische Religion*, 177; Morenz, *Egyptian Religion*, 168; Emily Teeter, "Maat," in *Oxford Encyclopedia of Ancient Egypt* (Oxford: Oxford University Press, 2001), 319–21.

16. Karenga, *Maat*, 32; on the "Mandate of Heaven," see below.

17. Morenz, *Ägyptische Religion*, 46–47; *Egyptian Religion*, 44–45; note that Conman, "It's About Time," 34–35 compares the Egyptian model of the heavens with the *Kai Tian* model of ancient China, on which see below.

18. Gathered and summarized by Wayne Horowitz, *Mesopotamian Cosmic Geography* (Winona Lake, IN: Eisenbrauns, 1998): 147–65; and Francesca Rochberg-Halton, "Mesopotamian

but an undifferentiated chaos of waters, both salt "Tiamat" and fresh "Apsu" waters.[19] This is described in one of the creation stories, the poem *Enūma elish*, which starts from chaos (mixed salt and fresh) waters:[20]

> When the heaven above had not been named,
> Earth below had not been called by name.
> Nothing but primeval Apsu, their begetter,
> Creative Tiamat, she who bore them all,
> They commingled waters as a single body.

After creation, there was a world continent that was surrounded by Ocean (*marratu*), apparently with lands across the Ocean stream, drawn on a map as triangles.[21] In the midst of the world continent rose Mount Mašu, where the sun rose and set, with its peak in the heaven and roots in the underworld. In other accounts, the cosmic Mēsu tree grew at the center, with its roots reaching down one hundred leagues into Apsu (the fresh waters below the earth) and its crown in the heaven of Anu.[22] Temples too served as pillars reaching from earth to heaven.[23] The heavens had gates, in the far east and far west, which the heavenly lights passed through when appearing or vanishing.[24] There were regions of heaven, among the lights and winds of the sky, named "Great Palace" (*Ešarra*) and "Great Shrine" (*Ešgalla*).[25] The heavens were sometimes said to be made of fresh water, sometimes of stone. The center of the world was a temple or a city,

Cosmology," in *Encyclopedia of Cosmology*, ed. Noriss S. Hetherington (New York: Garland, 1993); see also Schwabl, "Weltschöpfung," §43, cols. 1502–4.

19. Rochberg-Halton, "Mesopotamian Cosmology," 399–400.

20. Translation based on C. F. Whitley, "The Pattern of Creation in Genesis, Chapter 1," *JNES* 17 (1958): 32; and Horowitz, *Mesopotamian Cosmic Geography*, 107–9.

21. Wayne Horowitz, "The Babylonian Map of the World," *Iraq* 50 (1988): 147–65; and Horowitz, *Mesopotamian Cosmic Geography*, 20–42, dating the map to the 9th century BC or later; cf. also Horowitz, *Mesopotamian Cosmic Geography*, 318–62. Francesca Rochberg, "The Expression of Terrestrial and Celestial Order in Ancient Mesopotamia," in *Ancient Perspectives: Maps and Their Place in Mesopotamia, Egypt, Greece and Rome*, ed. Richard J. A. Talbert (Chicago, University of Chicago Press, 2012), 32–34, suggests for the map a date from the end of the eighth to the early seventh century BC.

22. Horowitz, *Mesopotamian Cosmic Geography*, 97 102 (Mašu), and 245, 326, 362 (Mēsu).

23. Rochberg-Halton, "Mesopotamian Cosmology," 402. Mircea Eliade, *Cosmologie și alchimie babiloniană* (Bucharest: Vremea, 1937), 32–35; trans. by Alain Paruit, as *Cosmologie et Alchimie Babiloniennes* (Paris: Gallimard, 1991), 41–43; Mircea Eliade, "Psychologie et Histoire des Religions—A propos du Symbolisme du «Centre»," *Eranos Jahrbuch* 19 (1950 [1951]): 247–282; repr. with revisions as *Images et Symboles* (Paris: Gallimard, 1952); trans. by Philip Mairet, as *Images and Symbols: Studies in Religious Symbolism* (New York 1961; repr. 1969), ch. 1, "Symbolisme du «Centre»," 33–72 / "Symbolism of the 'Centre'," 27–56.

24. Wolfgang Heimpel, "The Sun at Night and the Doors of Heaven in Babylonian Texts," *Journal of Cuneiform Studies* 38 (1986): 132–43.

25. Horowitz, "Babylonian Map," 107–49, especially 125–28.

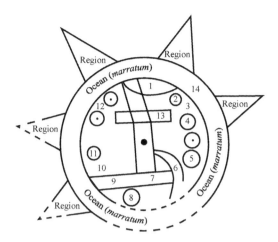

FIGURE 1.2. Babylonian World Map (BM 92687), courtesy Paul A. Whyman. 1. mountain (*ša-du-ú*), 2. city, 3. Urartu, 4. Assyria, 5. Der, 6. ??, 7. swamp (*ap-pa-ru*), 8. Susa. 9. channel (*bit-qu*), 10. Bit Yakin, 11. city, 12. Ḫabban, 13. Babylon, 14. Great Wall (*BÀD.GU.LA*)

because each of these sacred spaces symbolized "centeredness."[26] Out beyond civilized lands were wild nomadic peoples, barely human, who could invade and cause cosmic destruction.[27] The remote parts of the Earth were populated by marvelous beasts and products, and some heroes journeyed out to the edges of the Earth.[28] Overall, the Mesopotamian world-picture was schematic and ethnocentric, as one might expect for a culture operating according to traditional modes of thought,[29] and as seen in the "Babylonian World Map."[30]

26. Eliade, *Cosmologie şi alchimie babiloniană*, 25–32 ≈ *Cosmologie et Alchimie Babyloniennes*, 33–40; Mircea Eliade, *Le Mythe de l'Éternel Retour: Archétypes et Répétition* (Paris: Gallimard, 1949), 30–31, rev. ed. (1969), 24; ≈ Trans. by Willard R. Trask, as *The Myth of the Eternal Return* (New York: Harper, 1954; New York: Pantheon, 1954; London: Routledge & Kegan Paul, 1955; repr. New York: Harper, 1959, 1965; repr. Princeton, 2005), 12; Eliade, "Psychologie et Histoire des Religions," 49, 52–53 ≈ *Images and Symbols*, 39, 42; Rochberg-Halton, "Mesopotamian Cosmology," 402–3.

27. Eliade, *Le Mythe de l'Éternel Retour*, 25–29 ≈ *Le Mythe de l'Éternel Retour* (rev. ed.), 21–23 ≈ *The Myth of the Eternal Return*, 9–11; Eliade, "Psychologie et Histoire des Religions," 47–48 ≈ *Images and Symbols*, 37–38; Beate Pongratz-Leisten, "The Other and the Enemy in the Mesopotamian Conception of the World," in *Mythology and Mythologies: Methodological Approaches to Intercultural Influences*, ed. Robert Whiting (Helsinki: Neo-Assyrian Text Corpus Project, 2001), 195–231; Glenn M. Schwartz, "Pastoral Nomadism in Ancient Western Asia," in *Civilizations of the Ancient Near East*, ed. Jack M. Sasson (New York: Scribner, 2000), 1:249–58.

28. Mircea Eliade, *Le Mythe de l'Éternel Retour: Archétypes et Répétition* (Paris: Gallimard, 1949), as in note 27; Horowitz, *Mesopotamian Cosmic Geography*, 96–106.

29. Christopher R. Hallpike, *The Foundations of Primitive Thought* (Oxford: Clarendon, 1979); Jason Hawke, "Number and Numeracy in Early Greek Literature," *SyllClass* 19 (2009): 1–76.

30. Drawing based on Horowitz, *Mesopotamian Cosmic Geography*, 21–22, 402, pl. 2 (drawing), 406, pl. 6 (photo), and altered (cracks removed; locations and sizes of circles, triangles, and rectangles corrected), using a photo by the author and the photos published by the British Museum: http://www.britishmuseum.org/research/collection_online/collection_object_details.aspx?objectId =362000&partId=1.

India

The early Sanskrit texts known as the *Purāṇas*, an encyclopedic genre, record a world-picture that accords with the mytho-historical cosmology of India, and already before the *Purāṇas*, the epics present much of that material.[31] The epics often state that all things arose out of water, either an ocean or an undifferentiated mass; this action of rising from the water becomes the regular mode of creation expressed in the *Purāṇas*.[32] The epics describe the process of creation of the earth and heaven in varying ways, but when they describe creation as the separation of earth and sky, the beings responsible for the separation then set up a cosmic pole to keep the earth and sky apart.[33] The *Purāṇas* describe earth and sky as two bowls facing one another, with the sun traveling between them; the earth and sky are sometimes said to be the two wheels, bound by an axle, of the god Indra's chariot.[34] The axle seems to be analogous to the world-pillar of other traditions. In the epics, a single saltwater ocean surrounds the earth, but in the *Purāṇas*, there are seven concentric circular continents separated by oceans, only one being saltwater (and the others containing various fluids). In the epic texts, Meru is merely a high peak,[35] but in the *Purāṇas*, the cosmic mountain Meru is at the center of the innermost continent and plays a fundamental role in cosmology as the *axis mundi*.[36] The periodic destructions of the world always involve a massive flood, accompanied by earthquakes, heat, and wind.[37]

China

The Chinese texts offer several world-pictures, which share core features, especially that the *kosmos* evolves without any external divine agent and based only

---

Horowitz, *Mesopotamian Cosmic Geography*, 40–42 (see also Horowitz, "Babylonian Map," 154 ≈ Horowitz, *Mesopotamian Cosmic Geography*, 27), emphasizes the use of purely geometrical shapes, known from Babylonian geometry: the circles (*kippatu*) of the Ocean, concentric on their compass point, the triangle (*santakku*), the "river" shape (*nāru*), the long rectangle "brick mold" shape (*nalbattu*), and the "ox-eye" shape (*īni alpi*). The small circles for the cities also show central compass points, which we (author and artist) chose to retain.

31. Luis González-Reimann, "Cosmic Cycles, Cosmology, and Cosmography," in *Brill's Encyclopedia of Hinduism*, ed. Knut A. Jacobsen (Leiden: Brill, 2009), 2:411–28; Kim Plofker, "Humans, Demons, Gods and Their Worlds: The Sacred and Scientific Cosmologies of India," in *Geography and Ethnography: Perceptions of the World in Pre-Modern Societies*, ed. Kurt A. Raaflaub and Richard J. A. Talbert (West Sussex: Wiley, 2013), 32–42.

32. González-Reimann, "Cosmic Cycles," 412, 415–16.

33. González-Reimann, "Cosmic Cycles," 412.

34. González-Reimann, "Cosmic Cycles," 422–24.

35. E. Washburn Hopkins, "Mythological Aspects of Trees and Mountains in the Great Epic," *JAOS* 30 (1910): 366–74.

36. González-Reimann, "Cosmic Cycles," 424–27; Plofker, "Humans, Demons, Gods."

37. González-Reimann, "Cosmic Cycles," 415.

on its own intrinsic properties.[38] The primordial chaos, driven by the "inherent qualities of the Dao," produces order.[39] In Chinese cosmology, earth was always nearly flat (*ping*, 平). An early culture hero, the emperor Yü, is said to have sent out walkers in each cardinal direction, who reported the extent of the flat earth.[40]

In the earliest model, the flat earth was imagined to be shaped like the plastron (lower shell) of a tortoise, represented by the character *ya* 亞, which depicts four squares around a central square—that is, a quincunx pattern; the domed carapace of the tortoise represented the shape of the sky.[41] A second model, the Kai Tian ("Lid of Heaven," 盖天) model, explained sunset and sunrise by proposing that objects become invisible when more than 167,000 *li* away (there were about 400 meters to the *li*, 里). According to a third model, the Hun Tian ("enveloping heaven," 渾天) model, which originated in the first century AD, sunset and sunrise occur at same time all over earth, when the sun dips below, or rises above, the edge of the vaulted earth.[42] This measurement, similar to one made by Eratosthenes, in China was interpreted as showing how far the sun was above the flat earth, as shown in Figure 3.

After the world emerged from the primordial water, a cosmic flood nearly destroyed the *kosmos*, but a group of culture heroes ameliorated its effects.[43] The various Chinese models of the *kosmos* often included a cosmic tree (the Jian

38. John S. Major, "Myth, Cosmology, and the Origins of Chinese Science," *Journal of Chinese Philosophy* 5 (1978): 1–20; Joseph Needham, *Science and Civilization in China* (Cambridge: Cambridge University Press, 1959), 210–28; and overview by Xu Fengxian, "Astral Sciences in Ancient China," in *Oxford Handbook of Science and Medicine in the Classical World*, ed. Paul T. Keyser with John Scarborough (New York: Oxford University Press, 2018), 129–43.

39. John S. Major, *Heaven and Earth in Early Han Thought: Chapters Three, Four, and Five of the Huainanzi* (Albany: State University of New York, 1993), 23–25.

40. Lisa Raphals, "A 'Chinese Eratosthenes' Reconsidered: Chinese and Greek Calculations and Categories," *East Asian Science, Technology, and Medicine* 19 (2002): 10–60, 30. Similarly, Alexander the Great employed such walkers, or bematists, to measure his routes. Three men, named Philonides of Khersonesos (Keyser, *EANS* 659), Diognetos (Keyser, *EANS* 254), and Baiton (Keyser, *EANS* 186), are attested as a βηματιστής, see Pliny, *Nat. hist.* 2.181; 6.61–62; 6.69; 7.11; and 7.84, and Athenaios 10.59 (442c); Wilhelm Dittenberger and Karl Purgold, *Die Inschriften von Olympia*, Olympia 5 (Berlin: Asher, 1896) #276 (cols. 403–4).

41. Sarah Allan, *The Shape of the Turtle* (Albany: SUNY, 1991), 88–98.

42. Christopher Cullen, "A Chinese Eratosthenes of the Flat Earth: A Study of a Fragment of Cosmology in Huai Nan tzu 淮南子," *Bulletin of the School of Oriental and African Studies* 39, no. 1 (1976): 107–9; Major, *Heaven and Earth in Early Han Thought*, 23–53; Shigeru Nakayama, *A History of Japanese Astronomy: Chinese Background and Western Impact* (Cambridge: Harvard University Press, 1969), 24–43; Joseph Needham and Colin A. Ronan, "Chinese Cosmology," in *Encyclopedia of Cosmology*, ed. Noriss S. Hetherington (New York: Garland, 1993), 64–66.

43. Sarah Allan, *The Way of Water and Sprouts of Virtue* (Albany: SUNY, 2006), 39–41; Mark Edward Lewis, *The Flood Myths of Early China* (Albany: State University of New York Press, 2006), 28–33; Major, "Myth, Cosmology, and the Origins," 6; David W. Pankenier, "Heaven-Sent: Understanding Cosmic Disaster in Chinese Myth and History," in *Natural Catastrophes During Bronze Age Civilisations: Archaeological, Geological, Astronomical, and Cultural Perspectives*, ed. Benny J. Peiser et al. (Oxford: Archaeopress, 1998), 187–97.

FIGURE 1.3. Measuring the Height of the
Sun above the Flat Earth, courtesy
Paul A. Whyman.

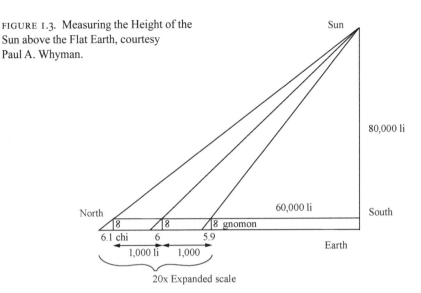

Tree) or a cosmic mountain named after the real Kunlun mountain but located in the remote west.[44] The cosmic mountain was the "Central Peak" (*Zhong Yue*, 中岳), the *axis mundi*, the place on earth closest to heaven, and a kind of paradise. Blessed persons were able to climb up to heaven on the cosmic tree. The earliest Chinese kingdom of which we have secure knowledge, the Shang, depicted the earth as square and aligned with the cardinal directions; they lived in the central area, which they named "Central Shang" (*Zhong Shang*, 中商).[45] Later Chinese empires reused the designation "central," usually in the name "Central Realm" (*Zhong Guo*, 中國).

All around this central realm lived barbarous peoples, destroyers of order, often analogized to animals, designated as the "Four Quadrates" (*Si Fang* 四方) and typically assigned to one of the four cardinal directions; among the emperor's duties was to banish or destroy them when necessary.[46] On the northern frontier,

44. Riccardo M. Fracasso, "Manifestazioni del simbolismo assiale nelle tradizione cinese antiche," *Numen* 28 (1981): 201–3; Mark Edward Lewis, *The Construction of Space in Early China* (Albany: State University of New York Press, 2006), 258–60; Major, *Heaven and Earth*, 26; David W. Pankenier, "The Cosmo-Political Background of Heaven's Mandate," *Early China* 20 (1995): 139–41. In the Zhou period, the world-mountain was "Mt. Sung 嵩山, which rises impressively from the yellow earth plain just southeast of Luoyang" (Pankenier, 139).

45. David N. Keightley, *The Ancestral Landscape: Time, Space, and Community in Late Shang China, ca. 1200–1045 B.C.* (Berkeley: University of California Press, 2000), 81; Pankenier, "Cosmo-Political Background," 139–41.

46. Allan, *The Shape of the Turtle*, 75–98; Lewis, *The Construction of Space*, 297.

they were typically nomads.[47] The emperor was obligated to perform "Heaven's Mandate" (*Tian Ming*, 天命) in order to guard the right functioning of the *kosmos* and, in particular, of the Central Kingdom, the land over which he ruled. All acts of the ruler affected the orderly operation of the *kosmos*, and its harmony would endure only so long as appropriate behavior persisted.[48] The emperor and his attendants conducted rituals that enacted the cycle of the seasons in the "Bright Shrine" or "Luminous Hall" (*Ming Tang*, 明堂), which imitated the structure of the *kosmos*.[49] Classical Chinese literature persistently promoted a "vital necessity of maintaining conformity with the normative patterns of the cosmos."[50]

Maya

The partially deciphered Mayan inscriptions and codices depict a world-picture that accords with the same mytho-historical cosmology evident in Egypt, Mesopotamia, India, and China. The world order prior to the current one was destroyed in a cosmic flood,[51] which wrought chaos in the *kosmos*.[52] The current creation began with an empty primordial sea beneath a dark primordial sky, and then the earth, beginning with the mountains, rose from the waters simply because the word of the creator gods called it forth.[53] Next, the "first father" (the Maize God) raised up the World Tree to support the sky, a support that was called "Raised-up-Sky" (*Wakah-Chan*),[54] and brought order out of primordial

47. Nicola DiCosmo, "The Northern Frontier in Pre-Imperial China," in *Cambridge History of Ancient China*, ed. Michael Loewe and Edward L. Shaughnessy (New York: Cambridge University Press, 1999), 885–966.

48. Lewis, *The Flood Myths*, 21–23; Geoffrey MacCormack, "Natural Law and Cosmic Harmony in Traditional Chinese Thought," *Ratio Juris* 2 (1989): 254–73; Victor H. Mair, *Wandering on the Way: Early Taoist Tales and Parables of Chuang Tzu* (New York: Bantam, 1994), 130–31; Needham and Ronan, "Chinese Cosmology," 63–64; Pankenier, "Cosmo-Political Background," 121–76; David W. Pankenier, *Astrology and Cosmology in Early China: Conforming Earth to Heaven* (Cambridge: Cambridge University Press, 2006), 226–34.

49. Lewis, *The Construction of Space*, 260–73; Pankenier, *Astrology and Cosmology*, 342–49.

50. Pankenier, *Astrology and Cosmology*, 317–50 (quotation on p. 350).

51. David Freidel, Linda Schele, and Joy Parker, *Maya Cosmos: Three Thousand Years on the Shaman's Path* (New York: Perennial, 1993), 105–6; Elizabeth Newsome, *Trees of Paradise and Pillars of the World: The Serial Stelae Cycle of '18-Rabbit-God K,' King of Copan* (Austin: University of Texas Press, 2001), 8–12, 198–207, 219. The early Inca myths included a flood narrative: Brian S. Bauer, Vania Smith-Oka, and Gabriel E. Cantarutti, eds., *Cristóbal de Molina: Account of the Fables and Rites of the Incas* (Austin: University of Texas Press, 2011), 4–13; Frank Salomon and George L. Urioste, *The Huarochiri Manuscript: A Testament of Ancient and Colonial Andean Religion* (Austin: University of Texas Press, 1991), 52–52 (§§29–34).

52. Newsome, *Trees of Paradise*, 198–205, 219.

53. Freidel, Schele, and Parker, *Maya Cosmos*, 59–60, 139–41, 283–84; Dennis Tedlock, *Popul Vuh: The Definitive Edition of the Mayan Book of the Dawn of Life* (New York: Touchstone, 1985), 72–75.

54. Freidel, Schele, and Parker, *Maya Cosmos*, 71–75; Newsome, *Trees of Paradise*, 198–207.

chaos.[55] Not much later, the Maize God was killed, but he rose again from the crack in a turtle carapace;[56] this turtle is identified as the asterism of the "Belt of Orion."[57] Maya myth often represents the earth itself as a quincunx diagram with the world-tree in the central square and a lesser tree in each cardinal square.[58] Sometimes the earth rests on the back of a turtle,[59] and the turtle was a model of the shape of the *kosmos*, with flat earth below and rounded heaven above.[60] The cardinal directions are oriented on the path of the sun, so that "north" and "south" are on one "side" or the other.[61]

The world-tree remained the *axis mundi*:[62] it both supported the world and provided a pathway for moving between levels of the *kosmos*.[63] It was often depicted as a cosmic maize plant.[64] In Maya art, the ruler of a city was depicted as the world-tree, or as the divine maize plant;[65] as such, he is also sometimes shown arising from a turtle-shaped altar.[66] The ruler of a city performed rituals to protect his subjects and ensure their success; through these rituals, the ruler also maintained the right order of the universe.[67]

Evidence from Hebrew

Although the cosmogonic story told in the Hebrew Bible is formally distinct from these accounts, the Hebrew texts include features that recall this mytho-historical cosmology.[68] At the divine word, creation emerges from watery chaos.

---

55. Linda Schele and David Freidel, *A Forest of Kings: The Untold Story of the Ancient Maya* (New York: Quill/Morrow, 1990), 255–58.

56. Freidel, Schele, and Parker, *Maya Cosmos*, 65, 92–94, 281–83, 370–72.

57. Freidel, Schele, and Parker, *Maya Cosmos*, 80–85.

58. Newsome, *Trees of Paradise*, 8–12, 218–19; Schele and Freidel, *A Forest of Kings*, 66; Evon Z. Vogt, *Tortillas for the Gods* (Cambridge: Harvard University Press, 1976), 13–16.

59. Schele and Freidel, *A Forest of Kings*, 66.

60. Karl A. Taube, "A Prehispanic Maya Katun Wheel," *Journal of Anthropological Research* 44 (1988): 193–99.

61. Vogt, *Tortillas*, 13–16.

62. Freidel, Schele, and Parker, *Maya Cosmos*, 53.

63. Lynn V. Foster, *Handbook to Life in the Ancient Maya World* (Oxford: Oxford University Press, 2005), 174–78.

64. Robert S. Carlsen and Martin Prechtel, "The Flowering of the Dead: An Interpretation of Highland Maya Culture," *Man*, n.s. 26, no. 1 (1991): 33–36.

65. Foster, *Handbook to Life*, 182; Freidel, Schele, and Parker, *Maya Cosmos*, 137, 394–97; Newsome, *Trees of Paradise*, 25–29, 129–30; Schele and Freidel, *A Forest of Kings*, 68.

66. Foster, *Handbook to Life*, 182.

67. Foster, *Handbook to Life*, 178–83; Kay A. Read, "Sacred Commoners: The Motion of Cosmic Powers in Mexican Rulership," *HR* 34 (1994): 39–69; Linda Schele and Mary Ellen Miller, *The Blood of Kings: Dynasty and Ritual in Maya Art* (New York: Braziller, 1986), 301–15.

68. Baruch Halpern, "The Assyrian Astronomy of Genesis 1 and the Birth of Milesian Philosophy," *Eretz-Israel* 27 (2003): 74–83, also draws this analogy, but only between Hebrew and Mesopotamian cosmogonies.

Genesis 1:2 uses language of "waste and empty" (וָבֹהוּ תֹהוּ, *tohū* and *bohū*), the "deep" (*təhōm*, תְהוֹם, translated by ἄβυσσος), and "the surface of the waters."[69] There are waters above the earth, presumably the source of rain (Gen 1:6), and waters below the earth, the source of springs (Exod 20:4; Deut 4:18; 5:8; 33:13 [*təhōm*]; Ezek 31:4 [*təhōm*]),[70] like the Mesopotamian Apsu or the Egyptian Duat. The sky above us, the "firmament" (רָקִיעַ *rāqîʿa*, translated as στερέωμα), is flat and shining, like a hammered sheet of metal (Gen 1:7–8, 14–15, 17, 20; Job 37:18).[71] The earth emerges from the waters (Gen 1:9), much as the first mound emerged from the waters in Egypt or as land emerges from receding floods anywhere.[72] The earth was imagined as a disk or "circle" (חוּג *ḥūg*, translated as γῦρος: Isa 40:22; cf. Prov 8:27), with the solid sky domed above (Job 22:14). Several locations are referred to as the "navel" of the world or the land (טַבּוּר *tabbūr*, translated as ὀμφαλός: Judg 9:36–37; Ezek 38:12).[73] Moreover, the original world order is destroyed in a cosmic flood (Gen 6:9–9:17), which arises from the original chaos—that is, from the abyss (*təhōm*, 7:11 and 8:2).

The concept of a world-pillar or world-mountain occurs in several places, but it mostly plays a minor cosmogonic role. First, pillars are part of the construction of the earth (Job 9:6; Ps 75:3) or even heaven (Job 26:11; עַמּוּד *ʿamūdh*, translated as στῦλος). Second, in Jacob's dream, a ladder connects earth and heaven (Gen 28:10–19). Third, with the divinity at its summit, Mount Sinai plays a central and

---

69. For *tohū* and *bohū*, see also Jer 4:23 and Job 26:7. See also Menahem Kister, "*Tohu wa-Bohu*, Primordial Elements and *Creatio ex Nihilo*," *Jewish Studies Quarterly* 14 (2007): 229–56. The chaos of the primal state is similar to Egyptian cosmogony: James E. Atwell, "An Egyptian Source for Genesis I," *JThS* 51 (2000): 451–53. The *təhōm* ("abyss") is etymologically related to the Babylonian Tiamat: James Albertson, "Genesis I and the Babylonian Creation Myth," *Thought: Fordham University Quarterly* 37 (1962): 230. Some of the following is noted by Wilhelm Schaefer, "Entwicklung der Ansichten des Alterthums über Gestalt und Grösse der Erde," in *Programm des Gymnasiums mit Realklassen zu Insterberg* (Insterberg: Carl Wilhelm, 1868), 4–5. For Hebrew words, see Wilhelm Gesenius, *Gesenius's Hebrew and Chaldee Lexicon to the Old Testament Scriptures*, trans. Simon P. Tregelles (New York: Wiley, 1893).

70. The watery "abyss" is also at Deut 33:13; Isa 51:10; Pss 77:16; 104:6; 135:6; Job 28:14; 38:16; 38:30; 41:31 (all of the sea). The sources of the Nile in Herodotus, *Hist.* 2.28 are from the abyss (i.e., Duat). Compare Albertson, "Genesis I," 230–31; Francesca Rochberg, "A Short History of the Waters of the Firmament," in *From the Banks of the Euphrates: Studies in Honor of Alice Louise Slotsky*, ed. Micah Ross (Winona Lake, IN: Eisenbrauns, 2008), 228–29.

71. The same root used of hammered metal objects is in Exod 39:3; Num 16:38–39; Isa 40:19; Jer 10:9. Compare Rochberg, "A Short History," 235–39.

72. Albertson, "Genesis I," 232–33.

73. Eliade, *Le Mythe de l'Éternel Retour*, 32–33 ≈ *Le Mythe de l'Éternel Retour* (rev. ed.), 25–26 ≈ *The Myth of the Eternal Return*, 13. Wilhelm H. Roscher, *Der Omphalosgedanke bei verschiedenen Völkern, besonders den semitischen: Ein Beitrag zur vergleichenden Religionswissenschaft, Volkskunde und Archäologie = Berichte über die Verhandlungen der Sächsischen Gesellschaft der Wissenschaften zu Leipzig: Philologisch-historische Klasse* 70.2 (Hildesheim: Olms, 1918), 12–25. The Hebrew word, however, might better be translated "high lookout site." Compare also Ezek 5:5 (Jerusalem is in the middle of all peoples and lands).

foundational role in the Jewish narrative (Exod 19); likewise, the divine epiphany was a pillar of fire by night and a pillar of cloud by day (Exod 13:21–22). Fourth, humans build the tower that would reach to heaven from earth (Gen 11:1–9).

All of this suggests that the early tellers of these tales were familiar with the mytho-historical cosmology, but they were selecting pieces from it for their own purposes.[74] There is no evidence that the behavior of the king maintained cosmic order or caused cosmic chaos.

The Cradle Cosmology

Thus, across these five ancient culture zones—Egypt, Mesopotamia, India, China, and Maya—there is a body of common belief in an early, mytho-historical, "cradle" cosmology. There are clear traces of it in the Hebrew texts, and, as we will see, this cosmology explains many of the neglected oddities of early Greek cosmologies. The key parallels among these five or six ancient culture zones are as follows.

1. The *kosmos* emerges from watery chaos, either spontaneously (China) or by divine fiat, perhaps by the word of the god(s) involved (Egypt, Israel, and Maya).

2. The *kosmos* has a center, and at that center is a world-tree, world-pillar, or world-mountain that serves an axis and passageway.

3. The earth is flat, either a disc (India, Israel, and Mesopotamia) or a quincunx (China, Maya, and perhaps Egypt).

4. The sky is solid, either flat (Egypt ⌐) or domed (China, Israel, and Maya).

5. The sky rotates around the earth; the changes in the sky are causally correlated with events upon the earth.

6. At the edge of the earth is the Ocean (the edges are underspecified in Egypt and China, but the watery Duat may be the Egyptian parallel to the outer Ocean); sometimes lands exist beyond that Ocean (Mesopotamia and India).

7. The *kosmos* is surrounded by chaos that threatens to invade, and (except in Egypt) a cosmic flood once almost destroyed the world.

8. The people who develop the myth are the only true people; around them are barbarians and monsters, who could invade.

9. The *kosmos* might redescend into chaos, so the ruler must act to maintain cosmic integrity (especially in China, Egypt, and Maya).

---

74. Indeed, the Phoenician cosmology of Sanchouniathon recorded by Philo of Byblos provides significant parallels to the Hebrew cosmology: see Uvo Hölscher, "Anaximander und die Anfänge der Philosophie (II)," *Hermes* 81 (1953): 392–97; M. L. West, "*Ab ovo*: Orpheus, Sanchuniathon, and the Origins of the Ionian World Model," *Classical Quarterly* 44 (1994): 289–307, at 295–302.

Earlier work by Eliade and others has shown the widespread existence of some of these features, in particular the key role of a "center" and the belief in a pillar or tree or mountain that ascends to heaven.[75] That humans, especially powerful ones like kings or emperors, might affect the operation of the kosmos is hardly mentioned by Greek texts. Instead, if they mention it at all, they allude to it for the purpose of rationalizing that idea. For example, in Sophocles's *Oedipus the King* 40–57, the king must heal the land, not by magic but by detecting the cause of the plague.

It is important to note that this "cradle" cosmology is one that "sublimates," so to speak, the political order of the so-called "palace" or "state" economy.[76] In China, "the realm of the supernatural was conceptualized by analogy with human socio-political experience"; then, the analogy was inverted, and the supernatural was understood as the cause of the human world.[77] That is, in a palace economy, most produce is gathered by the central ruler, who then redistributes it to his subjects, for whom he is obliged to provide. This was certainly the dominant economy in the empires of China, in pharaonic Egypt,[78] and also in the city-states of the Maya.[79] Such an economy is less well attested for Mesopotamia and India.

II. The Kozy Kosmos in Greece

The Kozy Kosmos was certainly found in a wide variety of cultures from an early date, some of which (Egypt and Mesopotamia) were in contact with Greece at the time. I am arguing that, in addition to the modern, the ancient atomist, and the Platonic-Aristotelian world-pictures, there was another, even earlier, world-picture, the Kozy Kosmos.[80] There is indeed reason to think that some of the

---

75. Roscher, *Der Omphalosgedanke*; Eliade *Le Mythe de l'Éternel Retour*; *The Myth of the Eternal Return*, as above §I.Mesopotamia; Giorgio De Santillana and Hertha von Dechend, *Hamlet's Mill: An Essay on Myth and the Frame of Time* (Boston: Godine, 1969), on which see Major, "Myth, Cosmology, and the Origins," 2–6.

76. Discussed by Michael E. Smith, "The Archaeology of Ancient State Economies," *Annual Review of Anthropology* 33 (2004): 73–102.

77. Pankenier, "Cosmo-Political Background," 161–76.

78. Brian P. Muhs, *The Ancient Egyptian Economy, 3000–30 BCE* (Cambridge: Cambridge University Press, 2016), suggests an increasingly mercantile economy starting in the first millennium BC, which was long after the formation of the mytho-historical cosmology.

79. C. Scott Speal, "The Evolution of Ancient Maya Exchange Systems: An Etymological Study of Economic Vocabulary in the Mayan Language Family," *Ancient Mesoamerica* 25 (2014): 69–113, notes that mercantilism appears only late in the Classic period, whereas terms for "the hierarchical appropriation of goods and labor" appear early and are pervasive.

80. My point of departure is similar to that of David J. Furley, *The Greek Cosmologists: The Formation of the Atomic Theory and Its Earliest Critics* (Cambridge: Cambridge University

elements of the Kozy Kosmos world-picture, and perhaps some version of the whole model, were known in Greek culture. I will argue that certain elements of the model were present. Furthermore, I will argue that many statements by early Greek writers make more sense if interpreted in light of the Kozy Kosmos model. There is, however, no extant extended description in Greek of the Kozy Kosmos model per se. But the elements of the model in other cultures are not gathered into one standard coherent account either. The distinction is that even early Greek texts seem to display debates and discussions as the Greeks transformed the model and invented radically new models. Nevertheless, as will be seen, many of the archaic elements persisted into much later periods.

My interpretation of these Greek texts as reflecting pieces of the Kozy Kosmos is a divergent interpretive move. The usual approach relies on interpretations of fragments and paraphrases of a limited set of early writers, which were supplied by ancient scholars committed to a teleological view of the development and ancestry of the Platonic-Aristotelian synthesis. But the context of these early writers was not the Platonic-Aristotelian synthesis. It is, rather, the reverse: Plato and Aristotle's context was the body of works of these earlier writers. Moreover, there is no *a priori* reason to think that Plato and Aristotle recorded precisely what their predecessors meant and, therefore, no reason to accept either author's interpretations, or those of their successors, rather than another system of interpretation. Therefore, I reject the hopelessly teleological term "pre-Socratic"—even if properly interpreted to include Aeschylus, Aristophanes, Arkhilokhos, Herodotus, Hesiod, and much of the Hippocratic corpus—because it conveys no content, only a scholarly attitude, and one that is actually wrong and maybe even dangerous.[81]

Center

Greeks referred to sacred places as an *omphalos* and described Greece as lying in a central, balanced, and optimal position on the earth.[82] Pindar records Delphic legends that claimed Zeus had sent two of his eagles from the utter west and east, who met at the *omphalos*.[83] The common Europe/Asia dichotomy—for example, in Herodotus, *Hist.* 1.4 and the Hippocratic *Airs, Waters, and Places*

Press, 1987), 1:1–2, who shows that the ancient atomist world-picture precedes the Platonic-Aristotelian world-picture.

81. All fragments cited from DK[6] or elsewhere are also cited with the actual source, to emphasize the degree to which we are dependent on doxographic sources whose motives were often far from accuracy.

82. Roscher, *Der Omphalosgedanke*, 61–78; James S. Romm, *The Edges of the Earth in Roman Thought* (Princeton: Princeton University Press, 1992), 65–66.

83. Pindar, *Pythians* 4.74, and *Paians* fr. 54 Maehler = Strabo, *Geogr.* 9.3.6 + Pausanias, *Descr.* 10.16.3; M. L. West, *The East Face of Helicon: West Asiatic Elements in Greek Poetry and Myth*

§§12–24 (Littré 2.52–92)—seems to reflect the same view, with central Greeks living at the western edge of Asia and the eastern edge of Europe.[84] Herodotus elsewhere accepts Libya as a third, coequal continent (4.42, 4.45), and Pindar may have also.[85]

Related to the view that "we" live in the "center" is the perception that east-to-west is the primary direction, as determined by the sun. The Egyptians posited an eastern sunrise peak, *Bachu*, matched by the western sunset peak, *Manu*,[86] and as noted above (§1), they regarded the central Nile as providing a cosmic directionality. Mesopotamian cosmic geography included a pair of solar mountains in east and west,[87] and the early Semitic context of the Hebrew scriptures also presupposes an east/west axis, so that the word for "north" means "left" and that for "south" means "right."[88] Early Greek epics seem to orient their world around the same east/west or dawn/dusk axis, in which they connect right and light with east and left and gloom with west, and place the paths of day and night adjacent (Homer, *Il.* 12.238–240; *Od.* 10.80–86, 190–192; 13.239–241; Hesiod, *Theog.* 746–757).[89]

(Oxford: Clarendon, 1997), 149–50. Epimenides explicitly denies an *omphalos*, proving that some believed in one, DK⁶ 3 B11 = Plutarch, *Failure of Oracles* §1 (409E).

84. The geographical work of Hekataios of Miletos (ca. 505 BC) had one book on Europe and one book on Asia, based on the citations of the fragments, primarily by Stephanos of Byzantium, *FGrHist* 1 F36–194 ("in 'Europe'") and F195–357 ("in 'Asia'"); see P. Kaplan, "Hekataios of Abdera," in *EANS* 361.

85. Pindar, *Nemeans* 4.70 (Europe), *Olympians* 7.18 (Asia), and *Pythians* 9.8 (Libya). The later geographical work of Eudoxos of Knidos (ca. 360 BC) in seven books treated Libya as a third continent: Friedrich Gisinger, *Die Erdbeschreibung des Eudoxos von Knidos* (Leipzig: Teubner, 1921), 12–18; François Lasserre, *Die Fragmente des Eudoxos von Knidos* (Berlin: de Gruyter, 1966), 240–41.

86. Eberhard Otto, "Bachu," *LÄ* 1 (1975): 574; Dieter Kurth, "Manu," *LÄ* 3 (1980): 1185–86.

87. Heimpel, "The Sun at Night," 143–46; Horowitz, *Mesopotamian Cosmic Geography*, 330–34.

88. Bernd Janowski, "Vom näturlich zum symbolischen Raum: Aspekte der Raumwahrnehmung in Alten Testament," in *Wahrnehmung und Erfassung geographischer Räume in der Antike*, ed. Michael Rathmann (Mainz: von Zabern, 2007), 51–64.

89. G. E. R. Lloyd, *Polarity and Analogy* (Cambridge: Cambridge University Press, 1966), 40, 47, cites anthropological studies of Indonesia and of the Nuer of the upper Nile. Graziano Arrighetti, "Cosmologia Mitica di Omero e Esiodo," *SIFC* 15 (1966): 1–60; Norman Austin, "The One and the Many in the Homeric Cosmos," *Arion* 1 (1973): 228–238; Norman Austin, *Archery at the Dark of the Moon* (Berkeley: University of California, 1975), 90–102; Alain Ballabriga, *Le soleil et le Tartare: L'image mythique du monde en Grèce archaïque* (Paris: Éditions de l'École des hautes études en sciences sociales, 1986), 60–62, 77, 108–17, cf. also 147–56, 175–255. Alain Ballabriga, *Les fictions d'Homère: L'invention mythologique et cosmographique dans l'Odyssée* (Paris: Presses universitaires de France, 1998), 7–10, 221–22, qualifies his earlier work by attributing certain geographical items in Homeric epic to actual geographical discoveries of the seventh and sixth centuries BC. See also G. W. Bowersock, "The East-West Orientation of Mediterranean Studies and the Meaning of North and South in Antiquity," in *Rethinking the Mediterranean*, ed. W. V. Harris (Oxford: Oxford University Press, 2005), 172–74.

Edges Extreme

As seen in all five cultures where the Kozy Kosmos is found (§I), the edges of the (flat) earth were in various ways extreme: inhabited by monsters or quasi-human beings. Already in the early poem of Aristeas of Prokonnesos, griffins who guarded gold and one-eyed paragons dwelt in the far north.[90] This picture of extreme edges is particularly apparent in Herodotus's *Histories*, but it is manifested in many Greek texts.[91] Herodotus exploited travelers' reports,[92] rather than the map of Anaximander (*Hist.* 4.42–44), and rejected the circumambient Ocean stream (2.23; 4.8, 36, 45). Instead, he portrayed most edges of the earth as desert land (*erēmos* or *erēmiē*), but all edges are extremes in what they produce (3.106.1).

Herodotus claimed that the far South was empty (4.185), although the nearer South produces huge elephants, tall, long-lived beautiful folk, and plentiful gold (3.114). In the furthest North, days and nights are said to be extreme, lasting six months each, straining even Herodotus's credulity (4.25). The far North was desert (4.17; 5.9), and beyond those limits one could not inquire (2.32; 4.31), but from the nearer North came much gold, perhaps guarded by griffins (3.116; 4.27). The far West is not described as empty. In Libya, the West was filled with things wild and rough, such as giant snakes and dog-headed men (4.191), and from far western Tartessos came extreme wealth for the first Greek who traded there (4.152). In the far East, where the sun burned close at dawn but was cool at evening (3.104), there was sand and more gold, excavated by giant ants (3.98; 3.102). Moreover, India was the farthest inhabited region of the East, and beyond that lay emptiness (4.40.2):

Μέχρι δὲ τῆς Ἰνδικῆς οἰκέεται Ἀσίη· τὸ δὲ ἀπὸ ταύτης ἔρημος ἤδη τὸ πρὸς τὴν ἠῶ, οὐδὲ ἔχει οὐδεὶς φράσαι οἶον δή τι ἐστί.

As far as India, Asia is inhabited: but from there on to the dawn, it is just desert, and no one can say what it's like.

Herodotus's known and inhabited world is symmetrical around its temperate central point, Greece (*Hist.* 3.106). Thus the Nile must resemble the Danube,

---

90. James D. P. Bolton, *Aristeas of Proconnesus* (Oxford: Clarendon, 1962), 74–103; Romm, *The Edges of the Earth*, 71–74; Tim Rood, "Mapping Spatial and Temporal Distance in Herodotus and Thucydides," in *New Worlds from Old Texts: Revisting Ancient Space and Place*, ed. Elton Barker et al. (Oxford: Oxford University Press, 2016), 115–16.

91. On Herodotus, see James S. Romm, "Dragons and Gold at the Ends of the Earth: A Folktale Motif Developed by Herodotus," *Merveilles & Contes* I (1987): 45–54; Romm, *The Edges of the Earth*, 9–41, 45–81, 83–94, 124–28, and 172–86.

92. O. A. W. Dilke, *Greek and Roman Maps* (Ithaca: Cornell University Press, 1985), 21–24, 57–59, and Romm, *The Edges of the Earth*, 32–41.

in mouths and course (2.33; 2.49), and if there were Hyperboreans living beyond the north wind, there would also be Hypernoteans living beyond the south wind (4.36).[93]

Roughly contemporaneous with Herodotus, and deploying similar notions about geographical symmetry, is a medical writer in the Hippocratic corpus, who sought to explain disease on the basis of directional exposure. He described the South as phlegmatic and moist, to which the North was symmetrically opposed, bilious and cold. The East was mild like spring, so that Asia was fertile, and the West was autumnal, while Greece being centrally located was moderate and best.[94]

From the same era are the reports of Skulax of Karuanda and of Ktesias of Knidos, each preserved only in extracts. From the eastern rim of the world, Skulax reported tribes whose physical characteristics stretched the limits of what could be called human, one-eyed beings or beings with feet or ears large enough to serve as parasols.[95] Ktesias reported on India, which had springs flowing with gold, rivers burgeoning with honey, and people who lived far beyond the short lifespans of the Greek world. But India also had monsters like the *martikhora* and the dog-headed people.[96]

Nearly contemporary with Ktesias, the historian Thucydides describes far western Sicily as extraordinarily large (*Hist.* 6.1.1), originally settled by monsters from the *Odyssey*, the Cyclopes and the Laestrygonians (6.2.1), and then filled with a plethora of strange barbarians (6.2.2–6), and altogether too extreme for conquest (6.6.1). Thucydides also attributes the origin of the Athenian plague, a hitherto unknown disease (2.47.3; 2.50.1), to the remotest reaches of the South, the epic land of "Ethiopia beyond Egypt" (2.48.1).[97]

This piece of the model did not wholly fade, but it is reprised by Strabo (around 20 AD), who defends the Homeric model of the inhabited world as

93. On symmetries in Herodotus's world-picture, see Klaus E. Müller, *Geschichte der antiken Ethnographie und ethnologischen Theoriebildung* (Wiesbaden: Steiner, 1972), 1:101–31; François Hartog, *The Mirror of Herodotus*, trans. Janet Lloyd (Berkeley: University of California Press, 1988), 14–19; James S. Romm, "Herodotus and Mythic Geography: The Case of the Hyperboreans," *TAPhA* 119 (1989): 97–113; and Romm, *The Edges of the Earth*, 60–61.

94. Hippocratic corpus, *Airs, Waters, and Places* §§12, 16, 23 (Littré 2.52–54, 2.62–64, 2.82–86); cf. Müller, *Geschichte der antiken Ethnographie*, 1:137–44; Jacques Jouanna, ed., *Airs, Eaux, Lieux* (Paris: Les Belles Lettres, 1996), 54–64; and J. Laskaris, "Hippokratic Corpus, Airs, Waters, Places," in *EANS* 406. Similar is *Regimen* 2.37–38 (Littré 6.528–534).

95. Cited in Tzetzes, *Khiliades* 7.629–636; Romm, *The Edges of the Earth*, 84–85; Kaplan, "Skulax of Karuanda," in *EANS* 745–746.

96. Mostly preserved as an epitome in Photios, *Library* §42 (pp. 45–50) = *FGrHist* 688 F45; Romm, *The Edges of the Earth*, 77–81, 85–88; Kaplan, "Ktēsias of Knidos," in *EANS* 496; Andrew Nichols, *Ctesias: On India* (London: Bristol Classical Press, 2011). The spring running gold, F45.9; the *martikhora*, F45.15 (Aristotle, *History of Animals* 2.1, 501a24–b1); the river of honey, F45.29; the long lives of the Indians, F45.32; and the dog-headed people, F45.37—on whom see Klaus Karttunen, "ΚΥΝΟΚΕΦΑΛΟΙ and ΚΥΝΑΜΟΛΓΟΙ in Classical Ethnography," *Arctos* 18 (1984): 31–36).

97. Cf. Rood, "Mapping Spatial and Temporal Distance," 112.

surrounded by Ocean (*Geogr.* 1.1–10).[98] Indeed, for Roman geographers, the northern Ocean became the symbol of the extreme edges of the world, and the ultimate West became again (or was still) the land of paradise.[99]

Creation Arises from Chaos

In the Kozy Kosmos model, creation commences from emptiness, chaos, or disorder. Creation, whether spontaneous or divine, is the imposition of order upon the unordered. Early Greek texts offer a similar picture, from Hesiod, Homer's *Iliad*, and Akousilaos, to Alkman, Anaxagoras, the Hippocratic *Fleshes*, and Pherekudes of Suros, and finally, Thales, Anaximander, and Anaximenes, on whom see below, §II.Milesians.[100]

Hesiod indeed starts from Chaos (ΧΑΟΣ) in *Theogony* 116, which is usually interpreted as a "gap," a lack of being, rather than as active disorder, although *Theogony* 736–745 seems to display a concept of chaos filled with more disorder (742–743: ἔνθα καὶ ἔνθα φέροι πρὸ θύελλα θυέλλῃ / ἀργαλέη).[101] The *Iliad*, in the "Deception of Zeus" (14.201; 14.245–246; 14.302), refers to Okeanos as the begetter of all things—apparently derived from Mesopotamian epic[102]—and elsewhere refers to Okeanos as the origin of all freshwater (21.195–197), parallel to the Apsu of Mesopotamia (or Duat of Egypt). Plato (*Theaitetos* 152e) interpreted these lines as declaring that:[103]

$$\pi\acute{\alpha}\nu\tau\alpha \ldots \ \check{\epsilon}\kappa\gamma o\nu\alpha \ \dot{\rho}o\tilde{\eta}\varsigma \ \tau\epsilon \ \kappa\alpha\grave{\iota} \ \kappa\iota\nu\acute{\eta}\sigma\epsilon\omega\varsigma$$

all things . . . (are) the offspring of flux and change

Plato (*Symposium* 178b) credits the idea of a primal Chaos to Akousilaos of Argos, whom scholars consider both historian and philosopher.[104] Pherekudes

---

98. Romm, *The Edges of the Earth*, 43.

99. Romm, *The Edges of the Earth*, 140–49, 156–71; Paul T. Keyser, "From Myth to Map: The Blessed Isles in the First Century BC," *AncW* 24 (1993): 149–68.

100. Friedrich Börtzler, "Zu den antiken Chaoskosmogonien," *Archiv für Religionswissenschaft* 28 (1930): 253–68.

101. G. S. Kirk, J. E. Raven, and M. Schofield, *The Presocratic Philosophers: A Critical History with a Selection of Texts*, 2nd ed. (Cambridge: Cambridge University Press, 1983), 36–41, stress the importance of the χάος in *Theog.* 736–745.

102. Walter Burkert, *The Orientalizing Revolution: Near Eastern Influence on Greek Culture in the Early Archaic Age*, trans. Margaret E. Pinder (Cambridge: Harvard University Press, 1992), 91–93; West, *The East Face of Helicon*, 144–48.

103. Kirk, Raven, and Schofield, *The Presocratic Philosophers*, 13–17.

104. DK⁶ 9 B2 = *FGrHist* 2 F6a; the claim is repeated in many later sources. Cf. Kirk, Raven, and Schofield, *The Presocratic Philosophers*, 18–19; Schwabl, "Weltschöpfung," §23, cols. 1464–66.

of Suros mentions Chaos and seems to equate it with primordial water.[105] Alkman, as preserved in a paraphrase, reports another cosmology that seems to start from primordial Chaos:[106]

τὴν ὕλην πάν[των τετα]ραγμένην καὶ ἀπόητον

the matter of all things was disturbed and unmade

Much later, Anaxagoras refers to an initial chaos-like state, in which all kinds of things were mixed in an undifferentiated mass. He writes:[107]

ὁμοῦ πάντα χρήματα ἦν, ἄπειρα καὶ πλῆθος καὶ σμικρότητα· καὶ γὰρ τὸ σμικρὸν ἄπειρον ἦν.

All things were together, unbounded in quantity and smallness: for the small was also unbounded.

Moreover, he seems to assume that the unbounded continues to exist outside the *kosmos* and surrounds it:[108]

ἀήρ τε καὶ αἰθὴρ ἀποκρίνονται ἀπὸ τοῦ πολλοῦ τοῦ περιέχοντος, καὶ τό γε περιέχον ἄπειρόν ἐστι τὸ πλῆθος.

---

105. DK⁶ 7 B1a = Achilles, *Isagoge* §3; Hermann S. Schibli, *Pherekydes of Syros* (Oxford: Clarendon, 1990), F64 (p. 164), and pp. 46–48; M. L. West, "Three Presocratic Cosmologies," *CQ* 13 (1963): 172.

106. Malcolm Davies, *Poetarum melicorum graecorum fragmenta* (Oxford: Clarendon, 1991), 51–52, fr. 5.2.iii = Kirk, Raven, and Schofield, *The Presocratic Philosophers*, 47–49 = D. L. Page, *Poetae melici Graeci* (Oxford: Clarendon Press, 1962; repr. 1967 with corrections), fr. 5 = *POxy* 24 (1957) #2390; see West, "Three Presocratic Cosmologies," 154–56.

107. DK⁶ 59 B1 = David Sider, *The Fragments of Anaxagoras*, 2nd ed. (Sankt Augustin: Academia Verlag, 2005), 68–76 = Kirk, Raven, and Schofield, *The Presocratic Philosophers*, fr. 467; the fragment is preserved in two pieces in Simplicius, *On Aristotle's 'Physics'* 1.4 (187a21) and (187b7); for the edition of Simplicius, see Hermann Diels, *Simplicii in Aristotelis Physicorum Libros Quattuor Priores Commentaria*, Commentaria in Aristotelem Graeca 9 (Berlin: Reimer, 1882), 155.23–30 and 164.13–20. See also DK⁶ 59 B4.b = Sider, *Fragments*, 102–7 = Kirk, Raven, and Schofield, *The Presocratic Philosophers*, fr. 468; the fragment is preserved in Simplicius, *On Aristotle's 'Physics'* 1.2 (184b15), Diels, *Simplicii in Aristotelis Physicorum Libros Quattuor Priores Commentaria*, 34.21–26. Cf. William K. C. Guthrie, *A History of Greek Philosophy 2: The Presocratic Tradition from Parmenides to Democritus* (Cambridge: Cambridge University Press, 1965), 294–304.

108. DK⁶ 59 B2 = Sider, *Fragments*, 77–81 = Kirk, Raven, and Schofield, *The Presocratic Philosophers*, fr. 488 = Simplicius, *On Aristotle's 'Physics'* 1.4 (187a21), Diels, *Simplicii in Aristotelis Physicorum Libros Quattuor Priores Commentaria*, 155.30–156.1. The verb ἀποκρίνομαι here seems to mean "secrete" as in medical texts, e.g., in the Hippocratic corpus: *Airs, Waters, Places* §9 (Littré 2.38.17); *Ancient Medicine* §14 (Littré 1.604.1); *Prognosis* §23 (Littré 2.178.13–14); *On Seed* §3 (Littré 7.474.5).

Air and Aither are being separated from the surrounding mass, and the surrounding is unbounded in quantity.

Probably around the same time, the Hippocratic *Fleshes* records the belief that the *kosmos* began with disordered matter (ὅτε ἐταράχθη πάντα ... ὅτε συνεταράχθη).[109]

For these writers on cosmology and cosmogony, whatever role it played in their text as a whole, the origin of the *kosmos* seems to have been some chaos-like state, which often is seen as continuing to exist in the outer Ocean.[110] Of course, the path from chaos to *kosmos* varied a great deal, and their works indeed often responded to, or disputed with, their predecessors, including lost works unknown to us.

Guardians of Order

The *kosmos* arose from Chaos, and might slip back into it; therefore, semi-divine agents must act to ensure that the *kosmos* endures. Hesiod declared the Chaos-born earth herself the guardian: Γαῖ' εὐρύστερνος, πάντων ἕδος ἀσφαλὲς αἰεὶ (*Theog.* 117). Anaximander, Herakleitos, Parmenides, and Anaxagoras referred to divine guardian(s) of order.

Anaximander indicates that balance is maintained:[111]

διδόναι γὰρ αὐτὰ δίκην καὶ τίσιν ἀλλήλοις τῆς ἀδικίας, κατὰ τὴν τοῦ χρόνου τάξιν

to render justice and recompense to one another for their injustice, according to the order of time

In other words, ministers of justice (order) discover the imbalance, and they pay recompense to restore justice (order). Scholars since Theophrastos have wanted to see here an implicit reference to the balance of ontological opposites like cold/hot,[112] but something like the protagonist's speech in Sophocles, *Aias* 646–648 + 669–676, seems more likely and less retrojective:

---

109. Hippocratic corpus, *Fleshes* §2–3 (Littré 8.584–586).

110. For the circumambient ocean as manifesting chaos, see Romm, *The Edges of the Earth*, 20–26; Schwabl, "Weltschöpfung," §46.b, col. 1510.

111. DK⁶ 12 B1 = Kirk, Raven, and Schofield, *The Presocratic Philosophers*, fr. 110 = Simplicius, *On Aristotle's 'Physics'* 1.2 (184b15), Diels, *Simplicii in Aristotelis Physicorum Libros Quattuor Priores Commentaria*, 24.13–21. Cf. Guthrie, *History of Greek Philosophy* 2:87–89.

112. Charles H. Kahn, *Anaximander and the Origins of Greek Cosmology* (New York: Columbia University Press, 1960), 166–96; Kirk, Raven, and Schofield, *The Presocratic Philosophers*, 118–121; Gregory Vlastos, "Equality and Justice in Early Greek Cosmologies," *CP* 42 (1947): 169–73, emphasizes the roles of justice and balance in early Greek cosmology, especially in Anaximander's model.

ἄπανθ᾿ ὁ μακρὸς κἀναρίθμητος
   χρόνος
φύει τ᾿ ἄδηλα καὶ φανέντα
   κρύπτεται·

...

καὶ γὰρ τὰ δεινὰ καὶ τὰ
   καρτερώτατα
τιμαῖς ὑπείκει· τοῦτο μὲν
   νιφοστιβεῖς
χειμῶνες ἐκχωροῦσιν εὐκάρπῳ
   θέρει·
ἐξίσταται δὲ νυκτὸς αἰανὴς κύκλος
τῇ λευκοπώλῳ φέγγος ἡμέρᾳ
   φλέγειν·
δεινῶν τ᾿ ἄημα πνευμάτων ἐκοίμισε
στένοντα πόντον· ἐν δ᾿ ὁ παγκρα-
   τὴς ὕπνος
λύει πεδήσας, οὐδ᾿ ἀεὶ λαβὼν ἔχει.

Long and countless time brings
   forth all unseen things, and hides
   all that appears:

...

The awesome and the mighty
   submit
to authority. So it is that snowy-
   wayed
winter gives place to fruitful
   summer;
night's dark orbit makes room for
white-horsed day to kindle her
   radiance;
the blast of dreadful winds allows
   the
groaning sea to rest; almighty
   Sleep, too,
releases his fetters, not holding
   forever.

Both Anaximander and Sophocles attribute authority to time, and Sophocles's verbal phrases τιμαῖς ὑπείκει (670), ἐκχωροῦσιν (671), ἐξίσταται (672), and ἐκοίμισε (674), correspond closely to the semantics of Anaximander's διδόναι ... δίκην καὶ τίσιν. It seems likely that Anaximander referred to his guardian as "governing" (κυβερνᾶν) the *kosmos*.[113]

Likewise, Herakleitos speaks of the Erinyes restoring the transgressive sun to his path:[114]

Ἥλιος γὰρ οὐχ ὑπερβήσεται μέτρα· εἰ δὲ μή, Ἐρινύες μιν Δίκης ἐπίκουροι
ἐξευρήσουσιν.

Sun will not overstep his measures; otherwise the Erinyes, guardians of Dike, will find him out.

---

113. DK⁶ 12 A15 = Kirk, Raven, and Schofield, *The Presocratic Philosophers*, fr. 108 = Aristotle, *Physics* 3.4 (203b10–13).

114. DK⁶ 22 B94 = Miroslav Marcovich, *Heraclitus* (Mérida, Venezuela: Los Andes University Press, 1967), fr. 52 = Kirk, Raven, and Schofield, *The Presocratic Philosophers*, fr. 226 = T. M. Robinson, *Heraclitus: Fragments* (Toronto: University of Toronto Press, 1987), fr. 94 (and p. 144) = Plutarch, *On Exile* 604A.

Similarly, Herakleitos's "thunderbolt" rules all,[115] and in Herakleitos's eventual *ekpyrosis*, the fire will judge and convict.[116]

Parmenides also has a guardian, the cybernetic goddess who regulates the appearing of the fires in the sky:[117]

δαίμων ἣ πάντα κυβερνᾶι / goddess ruling all things

Anaxagoras labels νόος (νοῦς) as boundless (ΑΠΕΙΡΟΝ) and places it over all things to organize them well:[118]

νοῦς δέ ἐστιν ἄπειρον καὶ αὐτοκρατὲς καὶ μέμεικται οὐδενὶ χρήματι, ἀλλὰ μόνος αὐτὸς ἐφ᾽ ἑωυτοῦ ἐστιν.

Mind is boundless and absolute and mixed with no thing, but is all alone by itself.

For the author of the Hippocratic *Fleshes*, it was the immortal "hot" that thinks, sees, and hears everything (ὃ καλέομεν θερμὸν, ἀθάνατόν τε εἶναι καὶ νοέειν πάντα καὶ ὁρῆν καὶ ἀκούειν καὶ εἰδέναι πάντα ἐόντα τε καὶ ἐσόμενα) and brought order out of chaos.[119]

115. DK⁶ 22 B64 = Marcovich, *Heraclitus*, fr. 79 = Kirk, Raven, and Schofield, *The Presocratic Philosophers*, fr. 220 = Hippolytus, *Refutation* 9.10.6, τὰ δὲ πάντα οἰακίζει Κεραυνός, (then Hippolytus interprets, probably accurately) τουτέστι κατευθύνει, κεραυνὸν τὸ πῦρ λέγων τὸ αἰώνιον. Compare also DK⁶ 22 B41 = Marcovich, *Heraclitus*, fr. 85 = Kirk, Raven, and Schofield, *The Presocratic Philosophers*, fr. 227 = Robinson, *Heraclitus: Fragments*, fr. 64 (and pp. 126–127) = Diogenes Laërtios, *Lives* 9.1.

116. DK⁶ 22 B66 = Marcovich, *Heraclitus*, fr. 82 = Robinson, *Heraclitus: Fragments*, fr. 66 (and p. 127) = Hippolytus, *Refutation* 9.10.6; see A. Finkelberg, "On Cosmogony and Eypyrosis in Heraclitus," *AJPh* 119 (1998): 195–222.

117. DK⁶ 28 B12, line 3 = Simplicius, *On Aristotle's 'Physics'* 1.2 (184b15), Diels, *Simplicii in Aristotelis Physicorum Libros Quattuor Priores Commentaria*, 39.12–20; see A. H. Coxon, *The Fragments of Parmenides*, revised and expanded edition, ed. Richard McKirahan and Malcolm Schofield (Las Vegas: Parmenides, 2009), 366–72, who explicitly compares Herakleitos's ruling Thunderbolt on p. 371.

118. DK⁶ 59 B12 = Sider, *Fragments*, 125–141 = Kirk, Raven, and Schofield, *The Presocratic Philosophers*, fr. 476 = Simplicius, *On Aristotle's 'Physics'* 1.4 (187a21), Diels, *Simplicii in Aristotelis Physicorum Libros Quattuor Priores Commentaria*, 156.13–16, and 1.4 (187b7), Diels, *Simplicii in Aristotelis Physicorum Libros Quattuor Priores Commentaria*, 164.24–25. Paraphrased by Plato, *Kratulos* 413c; cf Guthrie, *History of Greek Philosophy* 2:272–279.

119. Hippocratic corpus, *Fleshes* §2–3 (Littré 8.584–586).

World Mountains, Pillars, and Trees

World-mountains may have figured in early Greek thought; Olympos reaching to heaven would have been the last vestige.[120] There are traces in Greek texts of a world-pillar. Hesiod (*Theog.* 775–779) tells of terrible Styx, eldest daughter of Ocean, whose house is partly formed of heaven-high "silver pillars" (κίοσιν ἀργυρέοισι). Of Calypso's father Atlas, the *Odyssey* 1.53–54 says:

> . . . ἔχει δέ τε κίονας αὐτὸς
> μακράς, αἳ γαῖάν τε καὶ οὐρανὸν ἀμφὶς ἔχουσιν.

> . . . he himself holds the tall pillars
> that hold earth and heaven apart.

According to Pindar, *Pythian* 1.19–20, Mount Aetna is the snowy pillar that holds down Typhon:

> . . . κίων δ᾽ οὐρανία συνέχει, / νιφόεσσ᾽ Αἴτνα . . .

> . . . the heavenly pillar holds him down, / snowy Aetna . . .

In Aeschylus's *Prometheus Bound* 351–352, the protagonist explains that it is his brother Atlas himself who:[121]

> ἕστηκε κίον᾽ οὐρανοῦ τε καὶ χθονὸς / ὤμοις ἐρείδων

> stands bearing the pillar of heaven and earth / on his shoulders

Atlas is, so to speak, the personification of the world-pillar—he does *not* hold up a globe, from the outside, but a canopy, from the inside.[122]

---

120. Homer, *Il.* 15.187–193: Olympos is common to all parts of the *kosmos*; Homer, *Od.* 6.43–45: it has a blessed climate. See Johanna Schmidt, "Olympos," *RE* 18 (1939): 276–79; William Merritt Sale, "Homeric Olympus and Its Formulae," *AJPh* 105 (1984): 1–28, especially 1, where he states it is neither earth nor the sky, but "a third region whose nature and location are elusive," and 15, where he cites Homer, *Il.* 5.749–751 = 8.393–395, which appears to identify "heaven" and "Olympos."

121. See also Hesiod, *Theog.* 517–520, 746–748; and Ibukos, fr. 336, in Davies, *Poetarum melicorum graecorum fragmenta*, fr. 336, p. 303 = *Scholia to Apollonios of Rhodes*, 3.106 (τῶν τὸν οὐρανὸν κιόνων); cf. West, *The East Face of Helicon*, 148–49.

122. Edouard Tièche, "Atlas als Personifikation der Weltachse," *MH* 2 (1945): 165–86.

Moreover, there is an explicit cosmic tree attested by Pherekudes of Suros.[123] Two sources refer briefly to a tree, and on it a decorated cloak (φᾶρος) or robe (πέπλος). The cloak or robe was woven by the primordial god Ζάς for his bride-to-be, the primordial goddess Χθονίη, and embroidered with depictions of the earth (Γῆ) and Ocean (Ὠγηνός), that is, of the created *kosmos*.[124] The tree is described as "floating" or "flying" (literally, "winged," ὑπόπτερος), just like the bowl of the sun in Mimnermos,[125] and draped with the embroidered cloak (πεποικιλμένον φᾶρος).[126] This tree seems to be the beginning of a cosmogony like the Kozy Kosmos, in which the first act of creation was to raise the cosmic tree; the laying upon it of the cloak embroidered with earth and sea corresponds closely to the world-tree supporting the heaven and maintaining the separation of earth and sky.[127] Hebrew texts also speak of something being "stretched out" to form the heavens;[128] in two texts, the heavens are stretched like a "tent" or "curtain" (Isa 40:22: אֹהֶל / σκηνὴν / tent; Ps 104:2: רִיעָה / δέρριν / curtain).

Milesians (Thales, Anaximander, Anaximenes) Writing the Kozy Kosmos

From the point of view of the Kozy Kosmos model, it seems possible to clar-ify the contribution of the Milesian philosophers, with respect at least to their

123. On Pherekudes of Suros, see Kirk, Raven, and Schofield, *The Presocratic Philosophers*, 50–71; Schibli, *Pherekydes of Syros*; Schwabl, "Welfschöpfung," §22, cols. 1459–64; West, "Three Presocratic Cosmologies," 157–72.

124. DK⁶ 7 B2 = Kirk, Raven, and Schofield, *The Presocratic Philosophers*, fr. 53 = Bernard P. Grenfell and Arthur S. Hunt, *Greek Papyri, Series II: Classical Fragments and Other Greek and Latin Papyri* (Oxford: Clarendon, 1897), #XI, pp. 21–23; Schibli, *Pherekydes of Syros*, F68 (pp. 165–67). West, "Three Presocratic Cosmologies," 158 argues for the accentuation "Ζάς."

125. West, "Three Presocratic Cosmologies," 171n2, 172, also draws the parallel with Mimner-mos and argues that the tree was "flying" in Chaos; see also Schibli, *Pherekydes of Syros*, 73–74.

126. DK⁶ 7 B2 = Kirk, Raven, and Schofield (1983) fr. 55 = Schibli, *Pherekydes of Syros*, F69 (p. 167) = Isidoros in Clement, *Stromateis* 6.53.5, and DK⁶ 7 A11 = Kirk, Raven, and Schofield, *The Presocratic Philosophers*, fr. 56 = Schibli, *Pherekydes of Syros*, F73 (p. 168) = Maximus of Tyre 4 ("Who Knows Better about Gods, Poets or Philosophers?"), §4g; George Leonidas Koniaris, *Maximus Tyrius: Philosophumena—ΔΙΑΛΕΞΕΙΣ* (Berlin: de Gruyter, 1995), 45. For the ὑπόπτερος bowl of the sun in Mimnermos, *Nannō*, see Kirk, Raven, and Schofield, *The Presocratic Phi-losophers*, fr. 7; Bruno Gentili and Carolus Prato, *Poetarum elegiacorum testimonia et fragmenta*, 2nd ed. (Leipzig: Teubner, 2002) fr. 5 = Athenaios, *Deipnosophists* 11 (470a); see Archibald Allen, *The Fragments of Mimnermus* (Stuttgart: Steiner, 1993), 94–99. West, "Three Presocratic Cosmolo-gies," 167–69, denies that the cloak was put on the tree, but allows that cloak and tree both represent the *kosmos* somehow. Dirk L. Couprie, *Heaven and Earth in Ancient Greek Cosmology* (Berlin: Springer, 2011), 9–10 cites an Assyrian seal that appears to show a tree covered by a mantle.

127. Schibli, *Pherekydes of Syros*, 69–77.

128. In the Prophets: Isa 40:22 (נוֹטֶה / διατείνας); 42:5 (נוֹטֵיהֶם / [absent]); 44:24 (נֹטֶה / ἐξέτεινα); 45:12 (נָטוּ / ἐστερέωσα [sic]), 48:13 (טִפְּחָה / ἐστερέωσε [sic]); 51:13 (נוֹטֶה / ποιήσαντα [sic]); Jer 10:12 (נֹטֶה / ἐξέτεινε); 51:15 (נֹטֶה / ἐξέτεινε); Zech 12:1 (נֹטֶה / ἐκτείνων); and in the Writings: Ps 104:2 (נוֹטֶה / ἐκτείνων); and Job 9:8 (נֹטֶה / τανύσας).

theories about the origin of the *kosmos*. The primordial water of Thales is not an "element" in the sense of an irreducible substance from which, or with which, everything is made.[129] Rather, it is the same primordial unformed (ΑΠΕΙΡΟΝ) water that we have seen in the various versions of the Kozy Kosmos.[130] And just as the world that emerges from the primordial waters remains in some way surrounded by them, so too is Thales's earth "floating" (πλωτὴν) upon the water,[131] like the floating islands of *Odyssey* 10.3 (Αἴολος) and Herodotus, *Hist.* 2.156.1 (Χέμμις).

Anaximander posited ΑΠΕΙΡΟΝ in eternal motion as the origin of all things.[132] For millennia, scholars have given themselves endless headaches by attempting to explain how something so unlike an element can produce everything or be that from which, or with which, everything is made. In the context of the Kozy Kosmos, the "unbounded" is probably the primordial chaos, the unformed stuff out of which the whole world came and may resolve back into.[133] If Anaximander's ΑΠΕΙΡΟΝ is to be compared to any "physical" thing, the best choice might be Chinese *qi*, 氣.

---

129. DK⁶ 11 A14 = Kirk, Raven, and Schofield, *The Presocratic Philosophers*, fr. 84 = Aristotle, *On Heaven* 2.13 (294a28–31).

130. For Thales's water as ΑΠΕΙΡΟΝ, see Simplicius, *On Aristotle's 'Physics'* 3.4 (203a16), Diels, *Simplicii in Aristotelis Physicorum Libros Quattuor Priores Commentaria*, 458.23–29.

131. Hölscher, "Anaximander," 385–391; William K. Guthrie, *A History of Greek Philosophy 1: The Earlier Presocratics and the Pythagoreans* (Cambridge: Cambridge University Press, 1962), 54–61; Kirk, Raven, and Schofield, *The Presocratic Philosophers*, 88–95. All of these scholars draw connections with the Egyptian Nun and the Mesopotamian Apsu and Tiamat, as primordial water.

132. DK⁶ 12 A1 = Kirk, Raven, and Schofield, *The Presocratic Philosophers*, fr. 94 = Diogenes Laërtios, *Lives* 2.1. For "through the eternal motion" (διὰ τῆς ἀιδίου κινήσεως) of the ΑΠΕΙΡΟΝ, see DK⁶ 12 A9 = Kirk, Raven, and Schofield, *The Presocratic Philosophers*, fr. 119 = Simplicius, *On Aristotle's 'Physics'* 1.2 (184b15), Diels, *Simplicii in Aristotelis Physicorum Libros Quattuor Priores Commentaria*, 24.21–25, and DK⁶ 12 A11 = Kirk, Raven, and Schofield, *The Presocratic Philosophers*, fr. 101b = Hippolytus, *Refutation* 1.6.2. Cf. David J. Furley, *The Greek Cosmologists: The Formation of the Atomic Theory and Its Earliest Critics* (Cambridge: Cambridge University Press, 1987), 1:28–30.

133. Gregory Vlastos, Review of *Principium Sapientiae* by F. M. Cornford, *Gnomon* 27 (1955): 65–76, at 74–75, also connects Anaximander's ΑΠΕΙΡΟΝ with Hesiod's ΧΑΟΣ, and reads Anaximander as stating that reabsorption of the *kosmos* back into the ΑΠΕΙΡΟΝ is the resolution of the opposites. Cf. Leo Sweeney, *Infinity in the Presocratics: A Bibliographical and Philosophical Study* (Dordrecht: Springer, 1972), 5. Harold Cherniss, "The Characteristics and Effects of Presocratic Philosophy," *JHI* 12 (1951): 319–45, at 324–25, argues that Anaximander's ΑΠΕΙΡΟΝ is a "boundless expanse of indefinitely different ingredients . . . it is everything *in actuality*" (cf. Sweeney, *Infinity*, 8–9). Kahn, *Anaximander*, 231–39, explores early senses of the word ΑΠΕΙΡΟΝ, and concludes that it means "what cannot be passed over or traversed from end to end," and that for Anaximander it is "the living, divine force of natural change" (cf. Sweeney, *Infinity*, 24–26). Guthrie, *History of Greek Philosophy*, 1:83–87, suggests that ΑΠΕΙΡΟΝ was undifferentiated; similarly, Sweeney, *Infinity*, 55–65, "indeterminate." Kirk, Raven, and Schofield, *The Presocratic Philosophers*, 110, argue that the primary meaning of ΑΠΕΙΡΟΝ in Greek of the era must have been "spatially unlimited" but acknowledge "for Anaximander, the original world-forming stuff was indefinite, it resembled no one kind of matter." Romm, *The Edges of the Earth*, 9–13, emphasizes the sense "lack of boundary."

Anaximenes postulated that the origin of the *kosmos* was the perpetually moving and boundless air (ἀέρα ἄπειρον) and proposed an air-water-earth continuum to explain transformation.[134] In the context of the Kozy Kosmos, we can understand Anaximenes as simply saying that when the primordial fluid chaos "thins," it can become like the fires above the earth, and when it "thickens," it can become like wind, mist, water, and even earth. He was hardly referring to elements or substances, in Aristotle's sense, but he was simply referring to primordial chaos coalescing into stuff with which we are familiar. The entire *kosmos* was surrounded by this endless "air" (ὅλον τὸν κόσμον πνεῦμα καὶ ἀὴρ περιέχει).[135] Here also, the best physical correlate of Anaximenes's "air" might be Chinese *qi*, 氣, which would help explain how he made it divine.[136]

Responding to Chaos

We have seen how different cultures responded to the threat that primordial chaos might return to the *kosmos* by invading or irrupting (§I). Greek philosophy can be defined as the search for the stability that lies under the perceptible instability of the world. When such an outlook is confronted with the concept of a primordial chaos (disorder or formlessness) that might return, several responses are possible. One may embrace—that is, radically accept—the chaos and assert that the universe is fundamentally dynamic. Or, secondly, one may attempt to banish—that is, radically reject—the chaos and assert that the universe is fundamentally stable. Or, thirdly, one may attempt to strike a balance between stability and chaos. It is possible to view at least some of the early Greek writers as having adopted one of these three strategies.

Some writers embraced the chaos: first, Herakleitos, then Anaxagoras, and lastly Leukippos and Demokritos.

For Herakleitos, the universe was an eternal fire,[137] implying that dynamism was the underlying principle, presumably a dynamism derived from the pluripotent primordial chaos. His directive and generative "fire" was an agent of perpetual change but always cashed out in balanced "measures" (ἁπτόμενον μέτρα καὶ ἀποσβεννύμενον μέτρα, "ignited in measures, quenched in measures")[138] and enforced by Erinyes (DK[6] 22 B94).

134. DK[6] 13 A5 = Kirk, Raven, and Schofield, *The Presocratic Philosophers*, fr. 140 = Theophrastos in Simplicius, *On Aristotle's 'Physics'* 1.2 (184b15), Diels, *Simplicii in Aristotelis Physicorum Libros Quattuor Priores Commentaria*, 24.26–25.1, note κίνησιν … ἀΐδιον.

135. DK[6] 13 B2 = Kirk, Raven, and Schofield, *The Presocratic Philosophers*, fr. 160 = Aëtios 1.3.4. On Aëtios, see J. Meyer, "Aëtios," in *EANS* 37–38.

136. DK[6] 13 A10 = Kirk, Raven, and Schofield, *The Presocratic Philosophers*, fr. 145 = Aëtios 1.7.13.

137. Kirk, Raven, and Schofield, *The Presocratic Philosophers*, 197–200; Furley, *The Greek Cosmologists*, 33–36.

138. DK[6] 22 B30 = Marcovich, *Heraclitus*, fr. 51 = Kirk, Raven, and Schofield, *The Presocratic Philosophers*, fr. 217 = Clement, *Stromateis* 5.104.1. The analogy to money from DK[6] 22

Anaxagoras said that "seeds of everything were in everything," a claim from which have sprouted endless tangles for his interpreters.[139] But what if he simply meant that anything can emerge out of anything, just as if out of the the still underlying primordial chaos? He is quoted as saying that even in the current state of the world:[140]

οὐδὲ χωρὶς ἔστιν εἶναι, ἀλλὰ πάντα παντὸς μοῖραν μετέχει ... ἀλλ' ὅπωσ-
περ ἀρχὴν εἶναι καὶ νῦν πάντα ὁμοῦ. ἐν πᾶσι δὲ πολλὰ ἔνεστι

Nor can they exist apart, but all things have a share of all ... but as it was in the beginning also now all things are together. In all things there are many (things)

For Leukippos and Demokritos, the atoms were in perpetual motion, thus both embodying chaos (in the sense of disorder) and allowing for an unlimited number of combinations.[141] The atoms moved in an infinite void, and when they coalesced to form any one of the many worlds, it occurred by "cutting off" from the ΑΠΕΙΡΟΝ:[142]

τὸ μὲν πᾶν ἄπειρόν ... κόσμους τε ἐκ τούτου ἀπείρους εἶναι ... γίνεσθαι
δὲ τοὺς κόσμους οὕτω· φέρεσθαι κατὰ ἀποτομὴν ἐκ τῆς ἀπείρου πολλὰ
σώματα παντοῖα

B90 = Kirk, Raven, and Schofield, *The Presocratic Philosophers*, fr. 219 = Marcovich, *Heraclitus*, fr. 54 = Plutarch, *The 'E' at Delphi* §8 (388D). Cf. Guthrie, *History of Greek Philosophy*, 1:454–459.

139. DK⁶ 59 B1 and B4, as above, "Creation Arises from Chaos." See David J. Furley, "Anaxagoras in Response to Parmenides," in *New Essays in Plato and the Pre-Socratics*, ed. Roger A. Shiner and John King-Farlow, Canadian Journal of Philosophy Supplement 2 (Guelph, Ontario: Canadian Assocation for Publishing in Philosophy, 1976), 61–85, at 71–76; repr. in *Cosmic Problems: Essays on Greek and Roman Philosophy of Nature* (Cambridge: Cambridge University Press, 1989), 47–65, at 55–58. Compare the near-parallel in the Hippocratic *Diseases IIII* §3 (Littré 7.544–546), the earth contains innumerable distinct powers that nourish distinct plants.

140. DK⁶ 59 B6 = Sider, *Fragments*, 110–113 = Kirk, Raven, and Schofield, *The Presocratic Philosophers*, fr. 481 = Simplicius, *On Aristotle's 'Physics'* 1.4 (187b7), Diels, *Simplicii in Aristotelis Physicorum Libros Quattuor Priores Commentaria*, 164.26–165.1; and DK⁶ 59 B11 = Sider, *Fragments*, 123–24 = Kirk, Raven, and Schofield, *The Presocratic Philosophers*, fr. 482 = Simplicius, *On Aristotle's 'Physics'* 1.4 (187b7), Diels, *Simplicii in Aristotelis Physicorum Libros Quattuor Priores Commentaria*, 164.20–24.

141. DK⁶ 67 A6 = Kirk, Raven, and Schofield, *The Presocratic Philosophers*, fr. 555 = Aristotle, *Metaphysics* 1.4 (985b4–19), and DK⁶ 68 A37 = Kirk, Raven, and Schofield, *The Presocratic Philosophers*, fr. 556 = Aristotle in Simplicius, *On Aristotle's 'Heaven'* 1.10 (279b12), Johan L. Heiberg, *Simplicii in Aristotelis de Caelo Commentaria* (Berlin: Reimer, 1894), 295.1–9.

142. DK⁶ 67 A1.31 = Kirk, Raven, and Schofield, *The Presocratic Philosophers*, fr. 563 = Diogenes Laërtios, *Lives* 9.31. Cf. Guthrie, *History of Greek Philosophy*, 2:406–13; Furley, *The Greek Cosmologists*, 140–46.

The whole is unbounded ... and there are unbounded *kosmoi* from this (the whole) ... and the *kosmoi* come to be like this: many bodies of all sorts move by abscission from the unbounded

Demokritos could have conceived the void itself as the chaos-as-gap: the primordial emptiness persists throughout the *kosmos* as the gap between atoms, the void in which the atoms move.

Secondly, some writers attempted to banish the chaos and assert a radical underlying stability to the world. Parmenides denied anyone could think about "what is not,"[143] which seems to include thinking about chaos or the unbounded. Instead, the way of truth led to eternal stability, and a *kosmos* that in truth was perfect and complete, like a sphere.[144] Zeno responded similarly and raised paradoxes about change and plurality in order to argue that there is no real change at all.[145] Plato describes Zeno's aim as arguing for essential Oneness (*Parmenides* 127D–128C), and essential Oneness would support an eternal and unchanging *kosmos*. Epikharmos is credited with asserting the eternal world order and spoofing the idea that it emerged from Chaos.[146] Melissos argued for an eternal and changeless *kosmos* unbounded in extent:[147]

ἀεὶ ἦν ὅ τι ἦν καὶ ἀεὶ ἔσται.

It always was what it was and always will be.

---

143. DK^6 28 B2 = Kirk, Raven, and Schofield, *The Presocratic Philosophers*, fr. 291 = Proklos, *On Plato's Timaios* 2 (29c), Ernst Diehl, trans., *Procli Diadochi in Platonis Timaevm Commentaria* (Leipzig: Teubner, 1903), 1:345, and Simplicius, *On Aristotle's 'Physics'* 1.3 (186a24), Diels, *Simplicii in Aristotelis Physicorum Libros Quattuor Priores Commentaria*, 116.25–117.1; also DK^6 28 B6 = Kirk, Raven, and Schofield, *The Presocratic Philosophers*, fr. 293 = Simplicius, *On Aristotle's 'Physics'* 1.2 (185b5), Diels, *Simplicii in Aristotelis Physicorum Libros Quattuor Priores Commentaria*, 86.19–28, and 1.3 (186a24), Diels, *Simplicii in Aristotelis Physicorum Libros Quattuor Priores Commentaria*, 117.2–13. See Coxon, *Fragments*, 290–305.

144. Kirk, Raven, and Schofield, *The Presocratic Philosophers*, 248–54; especially DK^6 28 B8 = Kirk, Raven, and Schofield, *The Presocratic Philosophers*, frr. 295–296 = Simplicius, *On Aristotle's 'Physics'* 1.2 (185a20), Diels, *Simplicii in Aristotelis Physicorum Libros Quattuor Priores Commentaria*, 77.34–78.29, plus 1.3 (187a1), Diels, *Simplicii in Aristotelis Physicorum Libros Quattuor Priores Commentaria*, 144.22–146.26. See also below §III, on the spherical *kosmos*. Cf. Guthrie, *History of Greek Philosophy*, 2:43–49; Coxon, *Fragments*, 312–52.

145. Kirk, Raven, and Schofield, *The Presocratic Philosophers*, 263–69, 277–79.

146. DK^6 23 B1 = Lucia Rodríguez-Noriega Guillén, *Epicarmo de Siracusa: Testimonios y Fragmentos* (Oviedo: Universidad de Oviedo, Servicio de Publicaciones, 1996), fr. 248 (pp. 149–51) = Diogenes Laërtios, *Lives* 3.10 (from Alkimos, *Essays for Amyntas*: see Nancy Demand, "Epicharmus and Gorgias," *AJPh* 92 (1971): 453–63); on Epikharmos, see L. Rodríguez-Noriega, "Epikharmos of Surakousai," in *EANS* 291–92.

147. DK^6 30 B8 = Kirk, Raven, and Schofield, *The Presocratic Philosophers*, frr. 525 and 527 = Simplicius, *On Aristotle's 'Physics'* 1.2 (184b15), Diels, *Simplicii in Aristotelis Physicorum Libros Quattuor Priores Commentaria*, 162.23–26, and 1.2 (184b15), Diels, *Simplicii in Aristotelis Physicorum Libros Quattuor Priores Commentaria*, 109.29–32.

ἀλλ᾽ ὥσπερ ἔστιν ἀεί, οὕτω καὶ τὸ μέγεθος ἄπειρον ἀεὶ χρὴ εἶναι.

But since it exists always, so also it must always be unbounded in size.

All four of these authors, Parmenides, Zeno, Melissos, and Epikharmos, sought to assure readers that the *kosmos* is stable and eternal.

Thirdly, two poets of cosmogony can be read as attempting to depict a balance between unbounded chaos and bounded *kosmos*. Xenophanes's poetry provides evidence that he saw a cycle between *kosmos* and chaos (the unbounded).[148] He speaks of the upper bound of the earth pushing against the air, while its lower part reaches the ΑΠΕΙΡΟΝ:[149]

γαίης μὲν τόδε πεῖρας ἄνω παρὰ ποσσὶν ὁρᾶται
ἠέρι προσπλάζον, τὸ κάτω δ᾽ ἐς ἄπειρον ἱκνεῖται

This upper bound of the earth is seen at our feet,
touching the air, the lower comes to the boundless.

He seems to be speaking of the created *kosmos* (here, the earth) as existing in balance with the unbounded (chaos). Just as in Egyptian, Mesopotamian, and Mayan versions of the Kozy Kosmos, the primordial water continues to exist below the earth. Also out at the edges of the created *kosmos*, the unbounded chaos continues to lurk, and into it the sun vanishes daily:[150]

. . . τὸν ἥλιον εἰς ἄπειρον μὲν προϊέναι, δοκεῖν δὲ κυκλεῖσθαι διὰ τὴν ἀπόστασιν . . .

. . . the sun proceeds into the boundless, but appears to circle round because of the separation . . .

That is, when the sun reaches the outer chaos, the sun ceases to exist. The idea that a sufficiently distant sun would vanish is similar to the claim in the "Kai

148. Xenophanes's god may have been the guarantor of that orderly cycle. Cf. DK[6] 21 B25 = Kirk, Raven, and Schofield, *The Presocratic Philosophers*, fr. 171 = Simplicius, *On Aristotle's 'Physics'* 1.2 (184b15), Diels, *Simplicii in Aristotelis Physicorum Libros Quattuor Priores Commentaria*, 22.22–23.20, ἀλλ᾽ ἀπάνευθε πόνοιο νόου φρενὶ πάντα κραδαίνει ("but aloof from toil he trembles all things by the will of his mind"). Cf. J. H. Lesher, *Xenophanes of Colophon: Fragments; A Text and Translation with a Commentary* (Toronto: University of Toronto Press, 1992), 106–10.

149. DK[6] 21 B28 = Kirk, Raven, and Schofield, *The Presocratic Philosophers*, fr. 180 (also fr. 3) = Achilles, *Isagoge* §4. Compare Lesher, *Xenophanes of Colophon*, 128–31.

150. DK[6] 21 A41a = Kirk, Raven, and Schofield, *The Presocratic Philosophers*, fr. 179 = Aëtios 2.24.9.

Tian" theory of ancient China (see §I.China). Indeed, Xenophanes did say that the sun and stars are daily ignited at their rise, and daily quenched at their setting:[151]

... σβεννυμένους δὲ καθ᾽ ἑκάστην ἡμέραν ἀναζωπυρεῖν νύκτωρ καθά-περ τοὺς ἄνθρακας· τὰς γὰρ ἀνατολὰς καὶ τὰς δύσεις ἐξάψεις εἶναι καὶ σβέσεις . . .

... being quenched each day they are rekindled nightly like coals: so risings and settings are catchings and quenchings . . .

A third way in which Xenophanes reflects the Kozy Kosmos model, while representing the *kosmos* as in balance with the outer chaos, is to depict the cosmic flood as a periodic occurrence from which the *kosmos* regularly recovers:[152]

ἀναιρεῖσθαι δὲ τοὺς ἀνθρώπους πάντας, ὅταν ἡ γῆ κατενεχθεῖσα εἰς τὴν θάλασσαν πηλὸς γένηται, εἶτα πάλιν ἄρχεσθαι τῆς γενέσεως, καὶ ταύτην πᾶσι τοῖς κόσμοις γίνεσθαι μεταβολήν.

All people are wiped out, when the earth is carried down into the sea and becomes mud; then again there is the start of a creation (*genesis*), and this transition occurs in all *kosmoi*.

Xenophanes offers the evidence of fossil plants, molluscs, and fish, to support his claim that the earth is being dissolved into mud over time. Then, he connects that observation to the cosmic flood of the Kozy Kosmos model, but he asserts a cycle of flood and re-creation.[153]

---

151. Stars: DK⁶ 21 A38 = Aëtios 2.13.14. Sun: DK⁶ 21 A40 = Kirk, Raven, and Schofield, *The Presocratic Philosophers*, fr. 177 = Aëtios 2.20.3. Cf. Paul T. Keyser, "Xenophanes' Sun on Trojan Ida," *Mnemosyne* 45 (1992): 299–311.

152. DK⁶ 21 B33 = Kirk, Raven, and Schofield, *The Presocratic Philosophers*, fr. 184 = Hippolytus, *Refutation* 1.14.6.

153. The cosmic flood of the Kozy Kosmos appears elsewhere in Greco-Roman texts: Akousilaos DK⁶ 9 B20 = Clement, *Stromateis* 1.102; Iulius Africanus in Eusebius, *Preparation for the Gospel* 10.10.7; Plato, *Timaios* 22; *Kritias* 111–112; *Laws* 3 (677a); Hegesianax, *Phainomena*, cited in Hyginus, *Astronomia* 2.29 (on Hegesianax, see C. Cusset, "Hegesianax," in *EANS* 358); Apollodoros 1.7.2; Propertius 2.32.53–54; Ovid, *Metamorphoses* 1.151–415; Germanicus, *Aratea* 561–562; Manilius, *Astron.* 4.828–833; Lucan 1.653–654. Adrienne Mayor, *The First Fossil Hunters: Paleontology in Greek and Roman Times* (Princeton: Princeton University Press, 2000), 201–6, and *The First Fossil Hunters: Dinosaurs, Mammoths, and Myth in Greek and Roman Times* (Princeton: Princeton University Press, 2011), 210–11, discusses some of these texts, and the relevance of fossils, as in Xenophanes.

Empedokles attempts to balance the primordial chaos and the stability of the created *kosmos*. That is, his model allows for perpetual change, but with an underlying stability.[154] Unlike Thales's "water" or Herakleitos's "fire," which are not elements but cosmogonic principles, the four "roots" of Empedokles were pretty much "elements" in the sense later desired by Plato and Aristotle.[155] Can we perhaps understand Empedokles as having represented the primordial Chaos as "Strife," and the stability of the *kosmos* as "Love,"[156] analogous to the Egyptian Ma'at and the Chinese Mandate?

III. Rounding the Edges, Raising the Sky

We have seen, in the prior section, how the Kozy Kosmos, the mytho-historical "cradle cosmology" of a flat earth and central pillar, which arose from chaotic primordial waters, is reflected in many of the models and assumptions of early Greek writers. If we consider it within the context of the Kozy Kosmos, much of Greek cosmology becomes more explicable. Now, looking at the evolution of Greek cosmology itself, we should keep in mind that early Greek thinkers were not seeking a system similar to what Plato propounded or Aristotle advocated. Greek writers of the early centuries, up to the times of Plato and Aristotle, did not primarily ponder what elements explained the world, or what caused the *kosmos*, or even earth's shape or appearance when viewed from afar; instead, they inquired about the origin and nature of the *kosmos* and the earth within that *kosmos*. They speculated about the first origins of the world and offered models that explained day and night, month and season, and other phenomena of earth and heaven. Certain traditions continued to direct their debates, but few overarching or undergirding assumptions, doctrines, or goals restrained their hypotheses.

Yet the accounts of Aristotle, the doxographers, the neo-Platonist commentators, and most scholars since have depicted their discussion as a dialogue culminating in Peripatetic cosmology. Moreover, doxographers often perversely

---

154. Strife scatters the elements; love joins them into functional wholes: DK[6] 31 B27+31 = Kirk, Raven, and Schofield, *The Presocratic Philosophers*, fr. 358 = Simplicius, *On Aristotle's 'Physics'* 8.1 (252a5), Hermann Diels, *Simplicii in Aristotelis Physicorum Libros Quattuor Posteriores Commentaria*, Commentaria in Aristotelem Graeca 10 (Berlin: Reimer, 1895), 1183.23–1184.4; and DK[6] 31 B30 = Kirk, Raven, and Schofield, *The Presocratic Philosophers*, fr. 359 = Aristotle, *Metaphysics* 2.4 (1000b12–17).

155. DK[6] 31 B17.1–35 = Kirk, Raven, and Schofield, *The Presocratic Philosophers*, fr. 348+349 = Brad Inwood, *The Poem of Empedocles*, 2nd ed. (Toronto: University of Toronto Press, 2001) fr. 25 = Simplicius, *On Aristotle's 'Physics'* 1.4 (187a21), Diels, *Simplicii in Aristotelis Physicorum Libros Quattuor Priores Commentaria*, 157.25–159.8 (the whole fragment).

156. Kirk, Raven, and Schofield, *The Presocratic Philosophers*, 296–300.

retroject later syntheses onto the words of early philosophers.[157] In light of the Kozy Kosmos model, let us try to clear away some of the retrojected accretions, to see what sort of world-picture these early thinkers were proposing.

The early Greek *kosmos* was small and organically whole, with earth and heaven in close coherence. At first, this mytho-historical Kozy Kosmos had little room for questions such as the shape of the sky or the nocturnal location of the Sun. But philosophers propounded theories that developed and altered that picture, with the result that many competing views coexisted up to the era of Aristotle and after. While there was no single line of development, five pieces of the puzzle were shared among many of the theories: (1) solid sky; (2) high ridges at the rim of earth; (3) sun near earth at sunrise and sunset; (4) sun feeds on vapors from earth; and (5) sun and moon are small compared to the earth.

Solid Sky, Flat or Domed

First of these shared pieces was the notion that the sky was solid, like a roof over the earth. That is what we find in Pherekudes of Suros, for example,[158] and is the image behind the falling anvil of Hesiod, *Theogony* 722–745. This is also probably the meaning of the bronze or iron *ouranos* in Homer and Pindar.[159] For the early Egyptians, the heaven was flat (the flat sky-determinative, ⌐), whereas in the Hebrew scriptures, it seems to have been rounded. Some Greek texts seem to assume a flat heaven, on which the house of Zeus sat;[160] but Aristophanes represents both Hippon of Kroton and Meton of Athens as propounding a domed heaven, like the hemispherical πνιγεύς (*pnigeus*, "damper"), used in cooking.[161]

---

157. D. R. Dicks, *Early Greek Astronomy to Aristotle* (Ithaca: Cornell University Press, 1970), especially 41–42, advocates thorough reevaluation of early Greek astronomy. Catherine Osborne, *Rethinking Early Greek Philosophy: Hippolytus of Rome and the Presocratics* (London: Duckworth, and Ithaca: Cornell, 1970), especially 183–86, argues that the context of quotations in the doxographic source Hippolytus must be taken into account, for reading Herakleitos and Empedokles at least. Alan C. Bowen, *Simplicius on the Planets and Their Motions: In Defense of a Heresy* (Leiden: Brill, 2013), unpacks the retrojection in one crucial source, Simplicius.

158. Schibli, *Pherykydes of Syros*, 38–49. Cf. Homer, *Il.* 8.13, and Hesiod, *Theog.* 116–134.

159. Bronze heaven: Homer, *Il.* 17.425 χάλκεος οὐρανὸς; cf. 1.426; 14.173; the "very brazen heaven": *Il.* 5.504; *Od.* 3.2, οὐρανὸν ἐς πολύχαλκον (perhaps compare "bronze-rich" as in *Il.* 10.315; 18.289; *Od.* 15.425); and Theognis 868–869, in an *adunaton* (Ἔν μοι ἔπειτα πέσοι μέγας οὐρανὸς εὐρὺς ὕπερθεν / χάλκεος). The iron heaven, σιδήρεος οὐρανὸς: *Od.* 15.329 and 17.565. See West, *The East Face of Helicon*, 139–40; Couprie, *Heaven and Earth*, 10–11.

160. The "bronze-based house of Zeus" in: Homer, *Il.* 1.426; 14.173; 21.438; 21.505; *Od.* 8.321, χαλκοβατὲς δῶ(μ); cf. Pindar, *Nemeans* 6.3: ἀσφαλὲς αἰὲν ἔδος; *Pythians* 10.27.

161. Aristophanes, *Clouds* 95–96 = Hippon DK⁶ 38 A2: see Dover (1968), 106–7, and Jean-Claude Picot, "L'Image du ΠΝΙΓΕΥΣ dans les *Nuées*: Un Empédocle au charbon," in *Comédie et Philosophie: Socrate et les « Présocratiques » dans les Nuées d'Aristophane*, ed. André Laks and Rosella Saetta Cottone (Paris: Presses de l'École normale supérieure, 2013), 113–29. For Meton, see Aristophanes, *Birds* 1000–1001 (Meton not in DK⁶); R. E. Wycherley, "Aristophanes, *Birds*,

Since the early eleventh century AD, scientists have speculated and calculated that the apparent shape of the sky is flatter than a hemisphere;[162] such an illusion might have contributed to the belief in various cultures, including Greek, that the sky was to some degree domed. The doxographers, however, uncritically attribute the notion of a spherical heaven to most thinkers, probably solely on the basis that they speculated about the *kosmos*.

Rim Ridges

I turn next to the idea that the flat earth was fringed with skyscraping peaks behind which the sun hid at night. Herodotus described the Kaukasos in the east and the Atlas in the west as heaven-high (*Hist.* 1.203; 4.184). The author of the Hippocratic *Airs, Waters, Places* told the same of the northern Rhipaian peaks,[163] which Herodotus (4.25) referred to, but did not name, as lofty and inaccessible, and Aristeas of Prokonnesos claimed to have approached.[164] The doxographers record that Anaximenes and others explained night as the shadow of those rim ridges,[165] and Plato's myth in his *Phaidon*, 109–111, placed the

---

995–1009," *CQ* 31 (1937): 22–31, at 24–25; Nan Dunbar, ed., *Aristophanes, Birds; Edited with Introduction and Commentary* (Oxford: Clarendon, 1995), 555–56; H. Mendell, "Meton of Athens," in *EANS* 551–552. On the πνιγεύς vessel, see Brian A. Sparkes and Lucy Talcott, *Pots and Pans of Classical Athens* (Princeton: Princeton University Press, 1961), plate 36; Brian A. Sparkes, "The Greek Kitchen," *JHS* 82 (1962): 121–37, at 128, plate IV.2; Brian A. Sparkes and Lucy Talcott, *Black and Plain Pottery of the 6th, 5th and 4th Centuries B.C.*, Athenian Agora 12 (Princeton: Princeton University Press, 1970), #2021–2022: pp. 32, 233, 377, 397, fig. 19, and plate 97, "cooking bell."

162. The astronomer and mathematician "Alhazen," *Optics* 7.55 (ca. 1020 AD), suggested the sky has a flat shape, but later writers sought to establish that it was domed; see Robert Smith, *A Compleat System of Opticks* (Cambridge: Crownfield, 1738; repr. Bristol: Thoemmes Continuum, 2004), 1:63–67 (book I, §5); Wilhelm Filehne, "Die mathematische Ableitung der Form des scheinbaren Himmelsgewölbes," *Archiv für Physiologie: Physiologische Abteilung* 34, no. 1 (1912): 1–32; H. Dember and M. Uibe, "Über die scheinbare Gestalt des Himmelsgewölbes," *Annalen der Physik* 360, no. 5 (1918): 387–96; M. Luckiesh, "The Apparent Form of the Sky-Vault," *Journal of the Franklin Institute* 191 (1921): 259–63; Albert Miller and Hans Neuberger, "Investigations into the Apparent Shape of the Sky," *Bulletin of the American Meteorological Society* 26 (1945): 212–16; D. Ventateswara Rao, "Variation of the Apparent Shape of the Sky with Intensity of Illumination," *Current Science* 15, no. 2 (1946): 40–41; D. Ventateswara Rao, "Effect of Illumination on the Apparent Shape of the Sky," *Proceedings of the Indian Academy of Sciences, Section A* 25, no. 1 (1947): 34–42; Lloyd Kaufman and Irvin Rock, "The Moon Illusion, I," *Science* 136 (1962): 953–61, at 954–955; Helen E. Ross and George M. Ross, "Did Ptolemy Understand the Moon Illusion?," *Perception* 5 (1976): 377–85, at 381–85; J. T. Enright, "The Eye, the Brain, and the Size of the Moon: Toward a Unified Oculomotor Hypothesis for the Moon Illusion," in *The Moon Illusion*, ed. Maurice Hershenson (Hillsdale, NJ: Erlbaum, 1989), 59–122, at 110–13.

163. Hippocrates, *Airs, Waters, and Places* §19 (Littré 2.70).

164. Compare Damastes of Sigeion, *FGrHist* 5 F1 = Stephanos of Byzantium, s.v. Ὑπερβόρειοι. Bolton, *Aristeas*, 39–42 argues that Aristeas is the source of all these remarks.

165. Anaximenes DK⁶ 13 A7.6 + A14, compare Aristotle, *Meteorology* 2.1 (354 a29–30), as well as Anaxagoras DK⁶ 59 A7, Demokritos DK⁶ A94 = Aëtios 3.10.4–5, and Arkhelaos DK⁶ 60 A4.4 = Kirk, Raven, and Schofield, *The Presocratic Philosophers*, fr. 515 = Hippolytus, *Refutation* 1.9.4.

whole *oikoumenē* at the bottom of a deep hollow whose foothills were those same peaks, and whose rim lay up in the true sky, in a blessed land of beauty.[166] Aristotle, despite believing in a spherical earth, still reported astronomically tall peaks all around the Mediterranean: Kaukasos in the east, Rhipaians in the north, Purene in the west, and the Ethiopian mountains in the south.[167] Only when Dikaiarkhos brought the heights of mountains within the range of human *logos* by measuring their heights did this belief recede.[168]

Sun and Moon Close to Earth

Thirdly, the sun moving across the flat or domed sky was thought to be closer to the earth at dawn and dusk, a belief that may have originated in the optical illusion that sun and moon appear larger when near the horizon.[169] Herodotus (*Hist.* 3.104) and Ktesias described the sun over India as much closer to the earth than when over Greece.[170] The author of the Hippocratic *Airs, Waters, and Places* employed the belief to explain the climate of the far west and east.[171] Euripides reflects the same picture, in his *Phaithon*:[172]

... Μέροπι τῆσδ᾽ ἄνακτι γῆς,
ἣν ἐκ τεθρίππων ἁρμάτων πρώτην χθόνα
Ἥλιος ἀνίσχων χρυσέᾳ βάλλει φλογί·
καλοῦσι δ᾽ αὐτὴν γείτονες μελάμβροτοι

166. Romm, *The Edges of the Earth*, 125.

167. Aristotle, *Meteorology* 1.13 (350a18–b18).

168. Paul T. Keyser, "The Geographical Work of Dikaiarchos," in *Dicaearchus of Messana: Text, Translation, and Discussion*, ed. W. W. Fortenbaugh and Eckahrt Schütrumpf (New Brunswick, NJ: Transaction Publishers, 2001), 353–72.

169. Photography demonstrates that the effect is an illusion. The cause has been debated since antiquity: Eugen Reimann, "Die scheinbare Vergrößerung der Sonne und des Mondes am Horizont," *Zeitschrift für Psychologie* 30 (1902): 1–38, especially 1–2 for the ancient evidence; Kaufman and Rock, "The Moon Illusion, I," 956–61; Irvin Rock and Lloyd Kaufman, "The Moon Illusion, II," *Science* 136 (1962): 1023–31; Enright, "The Eye, the Brain, and the Size of the Moon"; Lloyd Kaufman and James H. Kaufman, "Explaining the Moon Illusion," *Proceedings of the National Academy of Sciences of the United States of America* 97 (2000): 500–505; Helen E. Ross, "Cleomedes (c. 1st century AD) on the Celestial Illusion, Atmospheric Enlargement, and Size-Distance Invariance," *Perception* 29 (2000): 863–71; and Lloyd Kaufman et al., "Perceptual Distance and the Moon Illusion," *Spatial Vision* 20 (2007): 155–75.

170. Ktesias, *Indika*, FGrHist 688 F45.12 and 45.18 = Photios, *Library* 72 (pp. 45b and 46a): the sun in India rose ten times the size it did in Greece.

171. Hippocratic corpus, *Airs, Waters, and Places* §§12, 16, 23 (Littré 2.52–54, 2.62–64, 2.82–86).

172. James Diggle, ed., *Euripides Phaethon; Edited with Prolegomena and Commentary* (Cambridge: Cambridge University Press, 1970), 55 (lines 1–7) and 78–83 (commentary) = Richard Kannicht, *Tragicorum Graecorum Fragmenta*, vol. 5 (Göttingen: Vandenhoeck & Ruprecht, 2004) fr. 771–772 = Strabo, *Geogr.* 1.2.27 (lines 1–5) + Stobaios 1.25.6 (line 6) + Vitruvius, *De architectura* 9.1.13 (line 7).

Ἔω φαεννὰς Ἡλίου θ' ἱπποστάσεις.
... θερμὴ δ' ἄνακτος φλὸξ ὑπερτέλλουσα γῆς
... καίει τὰ πόρσω τἀγγύθεν δ' εὔκρατ' ἔχει

... to Merops, the king of this country,
Which is the first land that, from his chariot and four,
As he rises, the Sun strikes with his golden flame:
The swarthy neighbors call it
Dawn's and Sun's bright stables.
... The lord's hot flame rising over the earth
... Burns what's far away, but keeps the nearer temperate.

The belief persisted into the era of Aristotle, as well as the later peripatetic *Problems*,[173] and is recorded by Strabo (*Geogr.* 3.1.4–5) from Poseidonios or Artemidoros around 100 BC.

Furthermore, an optical illusion may have contributed to the belief that the sun was closer to the earth at dawn and dusk: the illusion of diverging crepuscular rays. Rays of light shining through clouds from the setting or rising sun (or moon) seem to converge at the "vanishing point" for much the same reason as railroad tracks seem to converge; the vanishing point then seems at a finite and even close distance.[174] The geometry of Figure 3 (above) in which the distance from the flat earth up to the sun is measured seems to derive from that illusion.

Sun Nourished by Terrestrial Vapors

Furthermore, the fire of the sun was close enough to both warm the earth and be nourished by vapors arising from the earth. Herodotus said exactly that (*Hist.* 2.25), as apparently did Antiphon, *On Truth*:[175]

πῦρ ἐπινεμόμενον μὲν τὸν περὶ τὴν γῆν ὑγρὸν ἀέρα, ἀνατολὰς δὲ καὶ δύσεις ποιούμενον τῶι τὸν μὲν ἐπικαιόμενον ἀεὶ προλείπειν, τοῦ δ' ὑπονοτιζομένου πάλιν ἀντέχεσθαι.

173. Aristotle, *Meteorology* 2.5 (361 b36–2a7); pseudo-Aristotle, *Problems* 8.17, but compare *Problems* 25.5; 25.15.

174. David K. Lynch, "Optics of Sunbeams," *Journal of the Optical Society of America* A 4.3 (1987): 609–11; Janet Shields, "Sunbeams and Moonshine." *Optics and Photonics News* 5, no. 7 (1994): 57, 59.

175. DK[6] 87 B26 = Gerard J. Pendrick, *Antiphon the Sophist* (Cambridge: Cambridge University Press, 2002), Book 2, fr. 26 = Aëtios 2.20.15; see also G. J. Pendrick, "Antiphon of Athens," in *EANS* 99.

[The sun is] a fire that feeds on the moist air around the earth and produces its risings and settings by continually abandoning the scorched air and clinging in turn to air which is somewhat moistened.

The Hippocratic *Breaths* says the same thing:[176]

Ἀλλὰ μὴν ἡλίου τε καὶ σελήνης καὶ ἄστρων ὁδὸς διὰ τοῦ πνεύματός ἐστιν· τῷ γὰρ πυρὶ τὸ πνεῦμα τροφή· τοῦ δὲ πνεύματος τὸ πῦρ στερηθὲν οὐκ ἂν δύναιτο ζῆν· ὥστε καὶ τὸν τοῦ ἡλίου δρόμον ἀένναον ὁ ἀὴρ ἀέν- ναος καὶ λεπτὸς ἐὼν παρέχεται.

But the path of the sun and moon and stars is through the *pneuma*; for the nourishment of fire is *pneuma*. Fire deprived of *pneuma* cannot live, so the everlasting air, being light, provides the everlasting path of the sun.

The doxographers record the same from the writings of Xenophanes, Herakleitos, and others.[177] Instead of sunset being the shadow of a cosmically tall mountain range, Xenophanes and Herakleitos taught that it was the extinction of the solar fire, rekindled at dawn.[178]

The sun, moon, and stars, therefore, were moving fires, periodically hidden or extinguished, about whose shapes and sizes thinkers speculated variously. The doxographers tell us that Anaximenes believed the sun and moon to be flat like leaves.[179] Parmenides preferred a model in which they encircled the flat earth like wreaths (στεφάναι) or like wheel rims, the light shining through a hole.[180] Anaximander apparently said the same, and in neither case did the

---

176. Hippocratic corpus, *Breaths* §3 (Littré 6.94.13–17); for "δρόμον" William H. S. Jones, *Hippocrates*, vol. 2, LCL 148 (Cambridge: Harvard University Press, 1923) reads "βίον"—and brackets "ἀένναος καὶ" ("so the air being light provides the everlasting path of the sun"). Similarly, *Sacred Disease* §13 (Littré 6.386) says the south wind afflicts the sun and moon; *Regimen* 2.38 (Littré 6.530–532) says the sun drinks (ἐκπίνει) moisture from the earth.

177. Xenophanes DK[6] 21 A40 = Aëtios 2.20.3, DK[6] 21 A46 = Aëtios 3.4.4, on which see Keyser, "Xenophanes"; and Herakleitos DK[6] 22 A11 = Aristotle, *Meteorology* 2.2 (354 b33—355 a33) = Marcovich, *Heraclitus*, fr. 58, compare Diogenes Laërtios, *Lives* 9.9. The doxographers record the same also of Anaximenes DK[6] 13 A7.5 = Hippolytus, *Refutation* 1.7.5; Philolaus DK[6] 44 A18 = Aëtios 2.5.3; Demokritos DK[6] 68 B25 = Eustathios of Thessalonike, *Commentary on Odyssey* 12.62; and Mētrodoros of Chios DK[6] 70 A4 = pseudo-Plutarch, *Stromateis* 11.

178. Sunset: Xenophanes DK[6] 21 A41 = Aëtios 2.24.4, and see Keyser, "Xenophanes"; Herakleitos DK[6] 22 B6 = Marcovich, *Heraclitus*, fr. 58 = Robinson, *Heraclitus: Fragments*, fr. 6 (and p. 79) = Aristotle, *Meteorology* 2.2 (355a13).

179. Anaximenes DK[6] 13 A7.4 = Hippolytus, *Refutation* 1.7.4.

180. Wreaths: Parmenides DK[6] 28 A37 = Aëtios 2.7.1, and compare B12 = Simplicius, *On Aristotle's 'Physics'* 1.2 (184b15), Diels, *Simplicii in Aristotelis Physicorum Libros Quattuor Priores Commentaria*, 39.12–20.

wreaths require a spherical earth.[181] Herakleitos and Antiphon saw the sun and moon as bowls (σκάφαι) of fire, a view perhaps encouraged by the Egyptian Kozy Kosmos and the poetry of Mimnermos. The bowls were also used to explain the phases of the moon and solar and lunar eclipses; eclipses were seen as darkenings due to the bowls tilting away from the earth.[182]

Small Sun and Moon

The last of these shared notions that I will mention is the small size of the sun and moon. Anaxagoras is attested to have said the sun was about the size of the Peloponnesos, while Herakleitos thought it was "foot-sized."[183] The latter belief persisted into the thought of Epicurus, who defended the idea by arguing that the sun was in effect a special case, and did not shrink in appearance due to distance.[184]

The Kozy Kosmos in Greece was framed by a flat earth close beneath a flat sky, on or under which the fires of sun, moon, and stars moved, nourished by vapors arising from the earth. Three observations, later cited as evidence for a spherical earth theory by the doxographers, do not necessarily suggest a spherical earth, and are in fact compatible with a flat earth. The three observations that are compatible with the Kozy Kosmos model are: (1) star elevations, (2) solstices and equinoxes, and (3) elevations of the sun.

Compatible: Star Elevations

First, travelers moving far enough south or north can observe the elevations of stars changing, which Aristotle, *On Heaven* 2.14 (298a3–6), and later writers cited as evidence that the earth is spherical or at least curved. But that observation is equally compatible with a flat earth under a flat sky of moderate height,

---

181. Anaximander DK⁶ 12 B4 = Aëtios 2.20.1 (vent), A10 = pseudo-Plutarch, *Stromateis* 2 (bark, φλοιόν, around tree), and A21–22 = Achilles, *Isagoge* 19, and Aëtios 2.20.1; 2.21.1; 2.25.1. On his rings, see the reconstruction in Philip Thibodeau, "Anaximander's Model and the Measures of the Sun and Moon," *JHS* 137 (2017): 92–111. The wreaths sit in the plane of the ecliptic, and the central earth can be any shape.

182. Mimnermos, *Nannō*; Gentili and Prato, *Poetarum elegiacorum testimonia et fragmenta*, fr. 5 = Athenaios, *Deipnosophists* 11 (470a); Herakleitos DK⁶ 22 A1.9–11 = Marcovich, *Heraclitus*, fr. 61 = Diogenes Laërtios, *Lives* 9.9–11, and DK⁶ 22 A12 = Aëtios 2.24.3; 2.27.2; 2.28.6; 2.29.3; Antiphon, *On Truth*, DK⁶ 87 B28 = Pendrick, *Antiphon*, fr. 28 = Aëtios 2.29.3.

183. Anaxagoras DK⁶ 59 A1.8 = Diogenes Laërtios, *Lives* 2.8; compare A42.8 = Hippolytus, *Refutation* 1.8.8; Herakleitos DK⁶ 22 B3 = Marcovich, *Heraclitus*, fr. 57 = Robinson, *Heraclitus: Fragments*, fr. 3 (and pp. 77–78) = Aëtios 2.21, compare Diogenes Laërtios, *Lives* 9.7. Aëtios 2.21.1 claimed that Anaximander said the sun was the same size as the earth (DK⁶ 12 A21).

184. Epikouros, *Letter to Pythoklēs* §91; Lucretius 5.564–565; cf. Jonathan Barnes, "The Size of the Sun in Antiquity," *ACD* 25 (1989): 29–41.

as walking down a corridor with one's eye fixed on a ceiling fixture will show—that is, as you walk towards the ceiling fixture, in order to keep your eye on it, you must gradually raise your gaze until you are looking straight up. Similarly, the disappearance of southern stars while traveling moderately far northward (or their coming into view while going south) would easily be explained on a flat earth as due to phenomena on the horizon, such as haze or peaks.

Compatible: Observing Solstices and Equinoxes

Secondly, the observation of solstices and equinoxes is likewise compatible with a flat earth over which the sun seasonally oscillates south and north.[185] Herodotus for example explains solstices by the coldness or warmness of the air (*Hist.* 2.25–26), and the doxographers tell us that Anaxagoras and Diogenes of Apollonia said the same.[186] After solstices and equinoxes were explained in terms of a spherical cosmology, doxographers retrojected that cosmology onto any thinker who gave dates for solstices and equinoxes.

Compatible: Elevation of Sun from Two Places

The third observation cited by ancient doxographers as evidence of a spherical earth is the simultaneous measurement of the elevation of the sun in two places on the same meridian. Eratosthenes used this observation to determine the size of the earth, but his interpretation of his results depends on the theory that the earth is spherical and therefore cannot prove that theory. The same measurement was used, by one school of early Chinese astronomers, called Kai Tian ("Lid of Heaven"; see §I.China), to determine the distance from the flat earth to the sun, which is how Epicureans interpreted the results of Eratosthenes.[187]

Addition: The Kosmos Itself Is Spherical

Thinkers of the fifth century BC added three innovations to the existing picture of the Kozy Kosmos, not all of which immediately became standard, and none of which require changing the theorized size or shape of the earth. First was

185. For example, the observations recorded in the Hippocratic Corpus, *Airs, Waters, and Places* §11 (Littré 2.52), and the observations of solstices that are attested for Stonehenge, early Mesopotamia, early China, and many cultures for which a spherical earth is out of the question.

186. Aristotle, *Meteorology* 2.2 (354b33–355a33); Anaxagoras DK⁶ 59 A42 = Hippolytos, *Refutation* 1.8.9; Diogenes of Apollonia DK⁶ 64 A17 = Alexander of Aphrodisias, *On Aristotle's 'Meteorology'* 2.1 (353a32); Michael Hayduck, *Alexandri in Aristotelis Meteorologicorum Libros Commentaria* (Berlin: Reimer, 1899), 67.1–14.

187. G. E. R. Lloyd, *The Ambitions of Curiosity* (Cambridge: Cambridge University Press, 2002), 52–55.

the idea that the *kosmos* itself was spherical. Xenophanes and Parmenides both emphasized the essential unity of the *kosmos*, and Parmenides even compared it to a sphere, perfect and complete:[188]

αὐτὰρ ἐπεὶ πεῖρας πύματον, τετελε-σμένον ἐστί	Since, then, there is a furthest limit, it is completed,
πάντοθεν, εὐκύκλου σφαίρης ἐναλί-γκιον ὄγκωι,	From every direction like the bulk of a well-rounded sphere,
μεσσόθεν ἰσοπαλὲς πάντηι· τὸ γὰρ οὔτε τι μεῖζον	Everywhere from the center equally matched; for it must not
οὔτε τι βαιότερον πελέναι χρεόν ἐστι τῆι ἢ τῆι.	be any larger or any smaller here or there;
οὔτε γὰρ οὐκ ἐὸν ἔστι, τό κεν παύοι μιν ἱκνεῖσθαι	There is neither what-is-not, to stop it from reaching
εἰς ὁμόν, οὔτ' ἐὸν ἔστιν ὅπως εἴη κεν ἐόντος	Its like; nor is there a way in which what-is could be
τῆι μᾶλλον τῆι δ' ἧσσον, ἐπεὶ πᾶν ἐστιν ἄσυλον·	More here and less there, since it all inviolably is;
οἳ γὰρ πάντοθεν ἶσον, ὁμῶς ἐν πεί-ρασι κύρει.	For equal to itself from every direction, it lies uniformly within limits.

Scholars dispute whether Parmenides meant that the *kosmos* itself had the shape of a sphere, but I side with Gallop and Coxon in thinking that Parmenides would have regarded the question of the shape of the whole *kosmos* as meaningless and is here describing attributes of the *kosmos*, said to be perfect and boundless in the way that a sphere is.[189] There is no question of the earth itself being a sphere, as none of what Parmenides says would apply to the earth: the earth is indeed "more here and less there," especially in mountainous and maritime

188. DK[6] 28 B8.42–49 = Kirk, Raven, and Schofield, *The Presocratic Philosophers*, fr. 299 = Simplicius, *On Aristotle's 'Physics'* 1.3 (187a1), Diels, *Simplicii in Aristotelis Physicorum Libros Quattuor Priores Commentaria*, 146.15–22 (just these lines).

189. Dicks, *Early Greek Astronomy*, 51; David Gallop, *Parmenides of Elea: Text and Translation with Introduction* (Toronto: University of Toronto Press, 1984), 19–21, 98–100, notes that Eudemos (in Simplicius) said that "some" interpreted the reference to be to the shape of the whole *kosmos*. A. H. Coxon, *The Fragments of Parmenides* (Assen/Maastricht: Van Gorcum, 1986; rev. ed. by Richard McKirahan and Malcolm Scholfield, Las Vegas, Zurich, and Athens: Parmenides, 2009), 212–17 / 335–342, argues that Parmenides alludes to equilibrium, comparing Melissos's statement that being itself is nonphysical, Melissos DK[6] 30 B9 = Kirk, Raven, and Schofield, *The Presocratic Philosophers*, fr. 538 = Simplicius, *On Aristotle's 'Physics'* 1.2 (185b5), Diels, *Simplicii in Aristotelis Physicorum Libros Quattuor Priores Commentaria*, 87.5–7 and 1.3 (186a13), Diels, *Simplicii in Aristotelis Physicorum Libros Quattuor Priores Commentaria*, 109.34–110.2. See also Couprie, *Heaven and Earth*, 64–67.

Greece, and can scarcely be "equal to itself from every direction."[190] In any case, Empedokles asserted the spherical shape of the *kosmos*:[191]

(B27) οὕτως Ἁρμονίης πυκινῶι κρύφωι ἐστήρικται
Σφαῖρος κυκλοτερὴς μονίηι περιηγέι γαίων . . .
(B29) οὐ γὰρ ἀπὸ νώτοιο δύο κλάδοι ἀίσσονται,
οὐ πόδες, οὐ θοὰ γοῦν(α), οὐ μήδεα γεννήεντα . . .
(B28) ἀλλ᾽ ὅ γε πάντοθεν ἶσος <ἐοῖ> καὶ πάμπαν ἀπείρων
Σφαῖρος κυκλοτερὴς μονίηι περιηγέι γαίων

Thus it is fixed in the dense cover of harmony,
A rounded sphere, rejoicing in its joyous solitude . . .
For two branches do not dart from its back,
Nor feet nor swift knees nor potent genitals, . . .
But it indeed is <self->equal on all sides and totally unbounded,
A rounded sphere rejoicing in its surrounding solitude.

Some scholars read Aristotle as attesting that Anaxagoras said the same, but Anaxagoras clearly imagined a flat earth.[192] Aristotle's citation is not evidence that Anaxagoras believed in a spherical *kosmos*. Aristotle's manner of citation provides no evidence of Anaxagoras's beliefs; rather, it is another example of doxographic retrojection. Anaxagoras described upper and lower faces of his discoid earth, which Aristotle assimilated to his own model of a spherical earth.

Addition: The Vortex

Secondly, Empedokles and Anaxagoras also hypothesized the vortex (δίνη), the cosmic rotation that explains the ongoing rotation of the stars.[193] In a flat-earth,

190. The spherical earth is alleged for Parmenides by Diogenes Laërtios, *Lives* 9.21–22, who credits even Thales with such a theory, in what should be read as an astounding example of doxographic retrojection.

191. DK⁶ 31 B27–29 = Inwood, *The Poem of Empedocles*, frr. 33–34, with B27, B29 = Kirk, Raven, and Schofield, *The Presocratic Philosophers*, frr. 358, 357, respectively; B27 = Simplicius, *On Aristotle's 'Physics'* 8.1 (252a5), Diels, *Simplicii in Aristotelis Physicorum Libros Quattuor Posteriores Commentaria*, 1183.23–1184.4; and B28 = Stobaios, *Eclogae* 1.15.2; B29 = Hippolytus, *Refutation* 7.29.13.

192. Aristotle, *Meteorology* 2.7 (365a23–26) reports Anaxagoras's theory of quakes, using the words "above" and "below", which Aristotle then interprets in terms of a spherical earth. Anaxagoras, however, surely believed in a flat earth: DK⁶ 59 A42.3 = Theophrastos in Hippolytus, *Refutation* 1.8.3; cf. Couprie *Heaven and Earth*, 185–88.

193. Empedokles DK⁶ 31 A49a = Inwood, *The Poem of Empedocles*, fr. 40; cf. Aristotle, *On Heaven* 2.1 (284a24–26); DK⁶ 31 B35 = Inwood, *The Poem of Empedocles*, fr. 61 = Simplicius, *On Aristotle's 'On Heaven'* 2.13 (295a29), Heiberg, *Simplicii in Aristotelis de Caelo Commentaria*,

flat-sky *kosmos*, the downward fall of objects does not immediately demand explanation, but in the vortex according to Empedokles and Anaxagoras, heavy material collects at the center, a concept that Furley has insightfully called "centrifocal dynamics."[194] Leukippos and Demokritos, however, hypothesized a vortex without a spherical *kosmos* and accepted plural worlds, thus contravening centrifocal dynamics.[195] A spherical *kosmos*, the vortex, and centrifocal dynamics were all compatible with a low heaven close to the earth, and only Demokritos and Archelaos seem to have asserted the later commonplace that the earth was small compared to the *kosmos*.[196]

Discarding the Kozy Kosmos: The Spherical Earth

Late in the fifth century BC a Pythagorean, perhaps Philolaos, proposed that the planets and even the earth orbited in circles around a central fiery "hearth," and for the first time it was suggested that earth itself was spherical.[197] This was no more based on observation than was the spherical *kosmos*. Plato accepted, or at least mythologized about, a spherical earth, in the *Phaidon* and the *Timaios*, but the idea was ignored or rejected by the geographers Ktesias and Ephoros. Even Plato, in the *Phaidon*, describes the shape of the earth with sufficient ambiguity so that scholars have debated it extensively.[198] Aristotle was not the first to

---

528–30. Anaxagoras DK⁶ 59 B9, B12, on which see Sider, *Fragments*, 118–20, 125–41, who argues that Aristotle did not mean to attribute a vortex to Anaximenes, since *On Heaven* 2.13 (294b31–295a29) = DK⁶ 13 A67 attributed the vortex to "all," which includes Anaximenes if and only if it includes Pythagoreans and Anaximander.

194. Centrifocal dynamics: Empedokles DK⁶ 31 A49a = Inwood, *The Poem of Empedocles*, fr. 40, Anaxagoras DK⁶ 59 B15. See David J. Furley, "The Dynamics of the Earth: Anaximander, Plato, and the Centrifocal Theory" in *Cosmic Problems: Essays on Greek and Roman Philosophy of Nature* (Cambridge: Cambridge University Press, 1989), 14–26.

195. Leukippos's vortex in DK⁶ 67 A1.31 = Kirk, Raven, and Schofield, *The Presocratic Philosophers*, fr. 563 = Diogenes Laërtios, *Lives* 9.31. Demokritos's vortex in DK⁶ 68 A5 = Diodoros of Sicily, *Hist.* 1.7.1, and DK⁶ 68 A88 = Lucretius 5.621–636, plus DK⁶ 68 A1.45 = Diogenes Laërtios, *Lives* 9.45.

196. Demokritos DK⁶ 68 A1.31 = Kirk, Raven, and Schofield, *The Presocratic Philosophers*, fr. 563 = Diogenes Laërtios, *Lives* 9.31 (infinite number of *kosmoi*). Arkhelaos DK⁶ 60 A1 = Diogenes Laërtios, *Lives* 2.17 (μέγιστον τῶν ἄστρων τὸν ἥλιον, καὶ τὸ πᾶν ἄπειρον), and DK⁶ 60 A4.3 = Kirk, Raven, and Schofield, *The Presocratic Philosophers*, fr. 515 = Hippolytus, *Refutation* 1.9.3 (μέγιστον μὲν ἥλιον, δεύτερον δὲ σελήνην).

197. Carl A. Huffman, *Philolaus of Croton* (Cambridge: Cambridge University Press, 1993); Plato, *Phaidon* 108c, explicitly credits "someone," often assumed to be Philolaos. However, compare Aristotle, *On Heaven* 2.2 (284b6–286a1), on Pythagorean ideas about heavenly motion; DK⁶ 44 A16 = Aëtios 2.7.7 and A17 = Aëtios 3.11.3 refer to the central fire and the moving earth, but do not describe the shape of the earth.

198. J. S. Morrison, "The Shape of the Earth in Plato's *Phaedo*," *Phronesis* 4 (1959): 101–19; Furley, *The Greek Cosmologists*, 53–57.

propose a spherical earth, since he credits *mathematikoi* with estimating its circumference.[199]

Some geographers of the early fourth century BC rejected or ignored the spherical earth model. Ephoros of Kumē depicted a tabular world, rectangular and regular, with pillars at the four cardinal points, perhaps to hold up the sky:[200] pillars of Herakles to the west, a great Northern one,[201] those at the mouth of the Red Sea,[202] and probably those of Bacchos by the mouth of the Ganges.[203] This rectangular portrait is remarkably similar in its outlines to the Persian view of the world in the time of Darius I, according to which peoples lived at the corners of a rectangle whose center was Persia; the peoples ruled by Darius are Lydians to the northwest, Skythians to the northeast, Indians to the southeast, and the land of Kush to the southwest.[204] That is, a quincunx, as in the Kozy Kosmos. Even the geographer Eudoxos of Knidos (ca. 360 BC), despite his extensive interest in the spheres of the stars, is not reliably attested to have made the earth a sphere.[205]

The attraction of the spherical earth theory for Plato and Aristotle may have been that on a spherical earth, there is no edge whence chaos could invade. The eternal boundlessness of the surface of a sphere replaces the boundless chaos that lurks at the rim of the flat earth. I suggest there was a multistep evolution from the Greek version of the "Kozy Kosmos" to the spherical small earth *kosmos*:

1. Parmenides made an analogy between the perfection of a sphere and that of the changeless *kosmos*;

2. Empedokles made the whole *kosmos* spherical on theoretical grounds;

---

199. Value of forty myriad stades: Aristotle, *On Heaven* 2.14 (298a15–17).

200. For the pillars, compare above §2, the world-tree of Pherekudes of Suros.

201. According to Pausanias of Damaskos, *Periplous* 188–195, who listed his sources as Ephoros and more recent authors, *Periplous* 109–127; see also Auienius, *Ora Maritima* 88–89. The pillar is clearly distinct from the western ones and lies among the Kelts along the Ocean, but the text is uncertain; cf. Bianchetti (1990). Didier Marcotte, ed., *Géographes Grecs*, vol. 1, *Introduction générale; Ps.-Scymnos, Circuit de la Terre* (Paris: Les Belles Lettres, 2000), 164–66, rejects Höschel's emendation to βόρειος and suggests Βριάρεως, as in Aristotle fr. 678 Rose = Aelian, *Historical Miscellany* 5.3, and in Plutarch, *Failure of the Oracles* §18 (419e–420a), an allegedly old name for the pillars of Herakles.

202. Ephoros *FGrHist* 70 F172 = Pliny, *Nat. hist.* 6.199.

203. See Strabo, *Geogr.* 3.5.5; Pliny, *Nat. hist.* 6.49—not cited as Ephoros, but evidently a pre-Alexandrian source; cf. Diodoros of Sicily, *Hist.* 17.9.5; Arrian, *Anabasis* 5.29.1; Plutarch, *Alexander* 62.4; and Q. Curtius Rufus 9.3.19.

204. J. Wiesehöfer, "Ein König Erschießt und Imaginiert Sein Imperium: Persische Reichsordung und Persische Reichesbilder zur Zeit Darios I (522–486 v. Chr)," in *Wahrnehmung und Erfassung geographischer Räum in der Antike*, ed. Michael Rathmann (Mainz: von Zabern, 2007), 31–40.

205. Gisinger, *Die Erdbeschreibung*, 15–16, credits a spherical earth to Eudoxos, citing the anonymous report in Aristotle, *On Heaven* 2.14 (298a15–17) and the unreliable "ΤΕΧΝΗ ΕΥΔΟΞΟΥ" of ca. 180 BC. See Alan C. Bowen, "Papyrus Parisinus graecus 1," in *EANS* 622. Lasserre, *Die Fragmente*, does not credit a spherical earth to Eudoxos.

3. Empedokles hypothesized a vortex . . .
4. . . . which led him to introduce centrifocal dynamics;
5. Demokritos suggested that the earth was only a small part of the whole *kosmos*;
6. Philolaos rounded the corners of that small earth into a sphere and set it whirling around the "central fire";
7. Plato (ambiguously) adopted the spherical earth model of Philolaos, and then Aristotle formalized it with a causal explanation (earth falls to the "center").

The evolution was by no means inevitable, and not everyone followed all steps of the path: atomists accepted only the vortex and the small size of the earth, Plato and Aristotle rejected the vortex, and Philolaus displaced the earth from the center. Moreover, the relative order of the contributions of Demokritos and Philolaos is uncertain, and if Philolaos offered his model before Demokritos did, then clearly the spherical earth model did not immediately convince everyone. Plato's *Phaidon* only ambiguously adheres to a spherical earth theory, and the *Phaidon* is decades earlier than his *Timaios*, where sphericity seems far more explicit. On the other hand, a few decades after Aristotle, Epicurus chose to retain a flat Earth in order to allow ongoing chaos.

My story rejects much ancient doxography. But when Aristotle and other doxographers were reporting theories, they practiced a characteristic retrojection; that is, they applied their term for a concept or activity to the work of earlier thinkers attested to have written about an earlier form of the concept or activity. For example, from their point of view, any thinker speculating about the *kosmos* must have offered an opinion about its shape, about the shapes and sizes of the heavenly bodies, and so on. It was easy, all too easy, for doxographers to conform their sources to their own questions. We must doubt all the doxographers, including Aristotle.

IV. Vaster than Empires

Greek knowledge and perception of the earth and the peoples upon it evolved as they were confronted with new data about more and more remote lands and peoples. Much of what the Greeks wrote about remote places and peoples was based in part on evidence; for example, the north of Europe was and is colder and wetter than the Mediterranean littoral, and the peoples there are paler and taller. Two special cases display the anomalous persistence of aspects of the Kozy Kosmos model: (a) early Greek perceptions of the uttermost west; and (b) Christian topography and the work of Kosmas, "the Indian navigator."

Greek Occidentalism

Because evidence about the west accumulated more slowly, Greek perceptions about lands and peoples to their west tend to display a greater number of archaic mytho-historical features, and thus they confirm some features of the Greek version of the Kozy Kosmos model.[206] Greeks persistently schematized their world-picture, invoking notions of symmetry to impose order upon the data; the particular symmetries invoked changed over time, from an east-west axis, to an Ocean stream encircling the disk of the Earth, to a flat earth whose edges grew extreme, and thence to a spherical earth (§§II–III).

The early presence of such symmetries is a manifestation of a mode of thought common among societies operating according to traditional categories (as noted above, §I.Mesopotamia). The longer persistence of this mode among Greeks discussing the west shows how gradual the transition was from traditional to more open and evidential modes of thought, as discussed by Lloyd and others.[207] Societies operating according to traditional categories produce symmetric and schematized world-pictures: examples are known from numerous cultures, such as Polynesia,[208] the Semitic milieu of the Hebrew scriptures,[209] and the ancient Persians.[210]

When Egyptians gazed westward, they saw in sunset lands the kingdom of the happy dead.[211] In the heroic tales of the *Odyssey*, Proteus of the Sea prophesied Menelaus's translation to the Elysian Fields (*Od.* 4.563–568), across which ever pleasant Ocean breezes blew from the West, though surely no rhapsode knew or cared where lay Ogygia (*Od.* 5).[212] Hesiod in *The Catalogue of Women* told of the flight of the sons of the North Wind round the disk of the earth, passing over remote lands and peoples. His western landmarks were mythic tropes: "amber" (so perhaps the Eridanos), and the Laestrygonians (who attacked

---

206. An earlier presentation of "Greek Occidentalism" appeared in Paul T. Keyser, "Greek Geography of the Western Barbarians," in *The Barbarians of Ancient Europe: Realities and Interactions*, ed. Larissa Bonfante (Cambridge: Cambridge University Press, 2011), 37–70, at 38–45, followed by additional material on Greek views of western peoples and a discussion of Roman writers.

207. G. E. R. Lloyd, *Magic, Reason, and Experience* (Cambridge: Cambridge University Press, 1979), 226–67; Paul T. Keyser, "The Name and Nature of Science: Authorship in Social and Evolutionary Context," in *Writing Science: Medical and Mathematical Authorship in Ancient Greece*, ed. Markus Asper (Berlin: de Gruyter, 2013), 17–61.

208. Saul H. Riesenberg, "The Organization of Navigational Knowledge on Puluwat," in *Pacific Navigation and Voyaging*, ed. Ben R. Finney (Wellington, New Zealand: Polynesian Society, 1976), 91–128.

209. Janowski, "Vom natürlich zum symbolisch Raum."

210. Wiesehöfer, "Ein König."

211. Keyser, "From Myth to Map."

212. Compare Müller, *Geschichte der antiken Ethnographie*, 1:53–59.

Odysseus).[213] Tartessus in southwestern Iberia became known to Greeks sometimes as a prosperous land, sometimes as a city, in each case ruled by an aged happy king, with people who rejoiced in written laws.[214] Herodotus (*Hist.* 3.115) admits that he cannot speak with any certainty about the extreme parts of the west. Mimnermus of Smyrna sang of the sun's rays stored up at the edge of the Ocean—the same sun who asleep sails in his golden bowl around to the sunrise Ethiopians, starting from the sunset Hesperides.[215] But the Pillars of Herakles increasingly seemed a divine limit to westward navigation, beyond which one could not go.[216] They, or the Fountains of Night, marked the limit of the West. The other cardinal points too were primary—the North Wind, the Unfoldings of Heaven in the East, and the ancient Garden of Phoibos in the South:[217]

... ὑπέρ τε πόντον πάντ' ἐπ' ἔσχατα χθονὸς
νυκτός τε πηγὰς οὐρανοῦ τ' ἀναπτυχάς,
Φοίβου παλαιὸν κῆπον ...

... and over the whole sea to the ends of land,
the Fountains of Night, and the Unfoldings of Heaven,
(and) the ancient Garden of Phoibos ...

213. See Müller, *Geschichte der antiken Ethnographie*, 1:59–66. M. L. West, *The Hesiodic Catalogue of Women: Its Nature, Structure, and Origins* (Oxford: Clarendon, 1985), 84–85, 127–30, and 169–71, argues that the poem was probably later than Hesiod. The fragments of *The Catalogue of Women* are edited by Reinhold Merkelbach and M. L. West, *Hesiodi Theogonia, Opera det Dies, Scutum, Fragmenta Selecta*, 3rd ed. (Oxford: Clarendon, 1990). Frr. 151–57 cover the north, east, and south, mentioning Ethiopians, Libyans, Scythians, Pygmies, Hyperboreans, and the "Katoudaei" (cave-dwellers, i.e., Trogodytes?). Fragment 150 = *POxy* 11 (1915) #1358 mentions the western tropes; some editors restore AITNH (the peak on Sicily), perhaps thinking of Thucydides, *Hist.* 6.2.1 (Laestrygones with equally mythic Cyclopes on Sicily).

214. Stesikhoros: Page, *Poetae melici Graeci*, fr. 7 (p. 100) = Strabo, *Geogr.* 3.2.11 places Tartessus in the far west. Anacreon, in Page, *Poetae melici Graeci*, fr. 16 (p. 184) = Strabo, *Geogr.* 3.2.14, describes the king of Tartessus as very aged; Herodotus, *Hist.* 1.163, calls Tartessus a city, located in the far west, and still ruled by a very aged king (cf. 4.152); and Strabo, *Geogr.* 3.1.6, attributes versified laws written in the local script to the natives (probably following Polybius, or an even earlier source).

215. Mimnermos, *Nannō*; Gentili and Prato, *Poetarum elegiacorum testimonia et fragmenta*, fr. 10 = Strabo, *Geogr.* 1.2.40, and fr. 5 = Athenaios, *Deipnosophists* 11 (470a), on which see Allen, *Fragments*, 94–99; and Stesikhoros, Page, *Poetae melici Graeci*, fr. 8 (pp. 100–101) = Athenaios, *Deipnosophists* 11 (469e), the sun sails in a goblet, δέπας.

216. Pindar, *Isthmians* 4.11–13; *Nemeans* 3.21; 4.69–70; *Olympians* 3.43–44: on which see Thomas K. Hubbard, *The Pindaric Mind: A Study of Logical Structure in Early Greek Poetry* (Leiden: Brill, 1985), 11–27; Romm, *The Edges of the Earth*, 17–18; M. M. Willcock, ed., *Victory Odes: Olympians 2, 7, 11; Nemean 4; Isthmians 3, 4, 7 / Pindar* (Cambridge: Cambridge University Press, 1995), 75; in Euripides, *Hippolytus* 744–745, the "old man of the sea" bars the way; compare W. S. Barrett, *Euripides Hippolytos* (Oxford: Oxford University Press, 1964), 303–4.

217. Sophocles: Stefan Radt, *Tragicorum graecorum fragmenta*, 2nd ed. (Göttingen: Vandenhoeck & Ruprecht 1999), fr. 956 = Strabo, *Geogr.* 7.3.1.

An alternate western terminus was Erytheia, the isle of the red sunset.[218] The far western reaches, however marked, seemed the land of happiness, at least to the fugacious *choros* in Euripides's *Hippolytus* 732–751. The *Periplous* attributed to Skulax covers all of Iberia and Gaul in three brief sentences that mainly mention ports, and to Italy gives only thrice that much text; for him, the peoples out west are Iberes and Ligures, whereas the Kelts live only in Italy.[219] Another late-enduring belief was that the sun set into the Ocean waters, and that the consequent hissing could be heard: Strabo (*Geogr.* 3.1.5) records that Poseidonios refuted that claim, still being made by Artemidoros of Ephesos (ca. 100 BC), and Lucan (*On the Civil War* 9.624–625) seems to depict the same.

So the far west, a land of darkness as it were, was depicted as the home of marvels and monsters and the blessed dead. The further west was beyond the grasp of Greeks before Aristotle, despite the Greek settlements in Sicily and South Italy and Massalia. Notions of symmetry and extremes persisted longer in the depictions of the West than they did in other directions. But the far east was also a place where old views persisted.

Epicurean and Christian Topography from Pliny to Kosmas

Some Latin authors display a reluctance to accept a spherical earth, and some Christian writers took a similar, or more extreme, position, rejecting not only sphericity but attempting to recreate a model of the *kosmos* based on the Hebrew scriptures. Pliny (*Nat. hist.* 2.161–166) records that some people doubted the earth could be spherical, because things on the other side would fall off. But he is likely mocking Epicureans,[220] and in his own voice, Pliny insists that the earth is a sphere. Tacitus, perhaps influenced by contemporary Epicureanism, seems to have assumed that the earth was flat: in *Agricola* 12.4 (sun barely shadowed by low extremities of earth, *extrema et plana terrarum humili*), and in *Germany* 45.1 (the noise of the sun as it emerges from the sea—as in Strabo, *Geogr.* 3.1.5), he describes solar phenomena of the far North as if they are due to the edges of

---

218. Erytheia (or Erythreia) in the far west: Hesiod, *Theog.* 289–294; 982–983; Pherekudes of Athens, *FGrHist* 3 F18 = Strabo, *Geogr.* 3.5.4; Herodotus, *Hist.* 4.8.2; and Ephoros, *FGrHist* 70 F129 = Pausanias of Damascus (known as "pseudo-Skumnos"), *Periplous* 152–169. See Marcotte, *Géographes Grecs*, 52–55, on Pausanias's treatment of far western peoples, and 160–163 on Erytheia and environs; compare Daniela Dueck, "Pausanias of Damaskos," in *EANS* 630–31.

219. Pseudo-Skulax, *Periplous* §2–4 on Iberia and Gaul; §5 and §8–19 on Italy; compare P. Kaplan, "Skulax of Karuanda, pseudo," in *EANS* 746.

220. Compare Lucretius, *On the Nature of Things* 4.404–413: seas and lands lie between (*inter*) any nearby mountain peaks and the rising sun, as if on a flat earth. On the other hand, Frederik Bakker, *Epicurean Meteorology: Sources, Method, Scope and Organization* (Leiden: Brill, 2016), 162–263, argues that the Epicureans simply never rejected a flat earth, allowing it as one possible theory among many.

a flat earth (*Germany* 45.1: *illuc usque et fama uera tantum natura*).[221] Not much later, the Greek Christian apologist Theophilos of Antioch, in his *Defense to Autolukos* 2.13 (ca. 180 AD), offered a layered model of the universe with a flat earth,[222] based on texts treated above, §I, "Evidence from Hebrew."

Around 310 AD, the Latin Christian writer Lactantius (*Divine Institutes* 3.24) indeed rejects the antipodes, believing that things "over there" on the antipodes would fall off.[223] The theologian Diodoros of Tarsos (ca. 380 AD) denied that the heavens and earth were spherical, and like Theophilos before him, attempted a cosmology based solely on the Hebrew scriptures.[224] Augustine (*City of God* 16.9) conditionally grants that the earth may be a sphere (*etiamsi figura conglobata et rotunda mundus esse credatur siue aliqua ratione monstretur*), but he raises doubts about whether there might be land on the opposite side, and further, even if there be land there, whether it might be inhabited. Not much later, Theodoros of Mopsuestia (bishop from 392–428 AD) wrote in his *Commentary on Genesis* that the spherical earth model was merely a pagan supposition—probably his aim was rejecting the rotation of the earth (a theory sometimes attributed to Herakleides of Pontus).

Around 540 AD, the merchant and sailor Kosmas composed his *Christian Topography*, in which he argued that the earth was a flat expanse with a central mountain, around which the sun and moon and stars orbited, all enclosed in a cosmic box.[225] He explains that the walls of the *kosmos* are beyond the outer ocean, and the firmament (below heaven) is joined to them; the roof of the *kosmos* is like the *tholos* of a bath (4.8). Sunrises and sunsets are caused by the sun going behind the central cosmic mountain (4.11–12). The measurement of the angles of sunshadows he explains on the flat earth model, like the Epicureans and the early Chinese (6.1–7). His plan view of the flat earth has four quadrants around a central square from which the cosmic mountain rises, similar to the

---

221. Peter Steinmetz, "Tacitus und die Kugelgestalt der Erde," *Philologus* 111 (1967): 233–41; and Florian Mittenhuber, "Die Naturphänomene des hohen Nordens in den kleinen Schriften des Tacitus," *MH* 60 (2003): 44–59, argue that Tacitus refers to a spherical earth.

222. On Theophilos, see Andreas Kuelzer, "Byzantine Geography," in *Oxford Handbook of Science and Medicine in the Classical World*, ed. Paul T. Keyser with John Scarborough (New York: Oxford University Press, 2018), 932, who also mentions Diodoros and Theodoros.

223. On Lactantius, Augustine, and Theodoros, see Klaus Geus, "Der Widerstand gegen die Theorie von der Erde als Kugel: Paradigma einer Wissenschaftsfeindlichkeit in der heidnischen und christlichen Antike?," in *Exempla imitanda: Mit der Vergangenheit die Gegenwart bewältigen?*, ed. Monika Schuol et al. (Göttingen: Vandenhoeck & Ruprecht, 2016), 65–84, at 75–78; for Theodoros, we rely on a Syriac translation, itself rendered into German by Geus.

224. On Diodoros, see P. T. Keyser & G. L. Irby-Massie, "Diodoros of Tarsos," in *EANS* 249–50. We rely on Photios, *Library* §223, for a summary of his work.

225. Wanda Wolska-Conus, *Topographie chrétienne* (Paris: Cerf, 1968–1973); cf. N. Lozovsky, "Kosmas of Alexandria, Indikopleustes," in *EANS* 487; Geus, "Der Widerstand," 78–80; and Kuelzer, "Byzantine Geography," 933.

FIGURE 1.4. Kosmas's Kosmos, courtesy Paul A. Whyman.

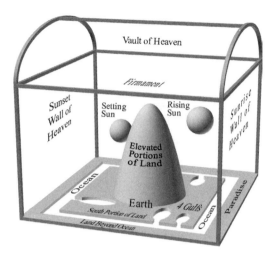

Chinese diagrams (6.34–35). One can see here the desire for a stable and orderly *kosmos*, safe from all possible chaos.[226]

V. Boundless Space, Endless Time

The mytho-historical "cradle" cosmology of a flat earth centered around "us" gave way—in Greece, in Greco-Roman cultures, and then in Medieval Europe, the Byzantine Empire, and the Islamic Caliphates—to a geocentric spherical earth model that was closely based on the arguments and theories of Plato and Aristotle. This model banished chaos: there was no edge through which chaos could irrupt, and the model, by asserting the divinely providential creation of the closed world, thereby asserted that the world was stable and perpetual.

In turn, the image of a geocentric spherical earth was discarded in favor of a heliocentric cosmology that gradually expanded, during the last five centuries, to allow for millions and billions of years and similarly ungraspable expanses of space. This model reintroduces instability and reveals the world, and ourselves within it, as contingent.

Something was lost in the transition from the Kozy Kosmos to the geocentric spherical earth model: the idea that humans could strongly affect the operation of the cosmos. The models of Plato and Aristotle posited a world that was

226. Figure 1.4 was created on the basis of copies of the manuscript drawings as published in Wolska-Conus, *Topographie chrétienne* (1928), 1:537, 543, 545, 549, 555, and especially 557, which the author translated and synthesized, and the artist then rendered visually perspicacious.

stable and operated on the basis of laws or principles that did not allow for human intervention. There was no chaos that could erupt, much less break in from "outside," since there was no outside at all. Therefore, there was no need for cosmic scale human agency—and not even the possibility. The Stoic model allowed for an external void and periodic conflagrations, but no human agency could affect those; likewise, the manifold crumbling worlds of the Epicureans admitted no prospect that humans could sustain them.

Ironically, as our *technē* has attained ever greater degrees of power, we have reached a point where we can in fact affect the earth as a whole. But we have no agreed framework, no mytho-history, that would explain or limit, much less enable, human intervention. Just as the mytho-historical Kozy Kosmos left traces and relics within later Greek speculations, so we are now faced with traces and relics of the now discarded models of Plato and Aristotle. Many of the specific observations that produce pressure on the modern synthesis were already being made in antiquity, when the dominant paradigm was the synthesis of Plato and Aristotle, the stable geocentric model.[227]

All in all, we find ourselves as yet unable to modify the agreed modern synthesis, so as to grasp that humans have gained sufficient power to alter, if not the whole *kosmos*, at least our pale blue sphere within it. The earth, whether central and motionless or peripheric and spinning, is now within our power to alter and not always for the better. The ancient mytho-historical Kozy Kosmos may yet have something to teach us, even if we dwell not on a flat earth with a central world-tree that ascends to heaven, but on a globe suspended in the void of heaven.

227. In an eventual expanded version of this paper, I hope to explore these considerations and sketch the gradually increasing pressure on the modern synthesis, which began almost before it was formed.

CHAPTER 2

Timosthenes of Rhodes

Duane W. Roller

TIMOSTHENES, WHO WAS ALLEGEDLY FROM RHODES but spent most of his career in Alexandria, is remembered as the author of *On Harbors*, a nautical guide for Ptolemaic seamen. He served under Ptolemy II (ruled 282–246 BC), and wrote his ten-book treatise sometime during that reign or shortly thereafter. It survives today in only thirty-eight fragments. In the late Hellenistic period the work was considered authoritative, especially for its distances, and was quoted by the major geographical authors of the era, beginning with Eratosthenes, a generation or two after Timosthenes, and including Hipparchos, Polybios, Strabo, Pliny, and others.

As far as is known, his professional career was totally associated with Alexandria and the government of Ptolemy II. He was perhaps born near the end of the fourth century BC and may have survived into the reign of Ptolemy III. The only information about his background or education is that a single source reports he was from Rhodes (F1), and some of his data may reflect residency there (F6). Timosthenes entered the royal service sometime after the accession of Ptolemy II in 282 BC, presumably one of those who came to Alexandria with the founding of the Library by the king.

The three sources that mention his career all make it clear that he held an important position in the Ptolemaic naval service, but each varies as to his exact title. The earliest record is by Strabo, who called him *nauarchos* ("naval commander"), essentially the naval chief of staff, an office created by Ptolemy I perhaps in 286 BC (*Geogr.* 9.3.10).[1] The other extant Greek term for Timosthenes's position, *archikybernetes* ("chief pilot"), cited by Markianos (*Epitome of the Periplous of Menippos* 2), recalls Onesikritos, who called himself by that term when navigating the fleet commissioned by Alexander the Great to sail from the Indos to the Persian Gulf in 325–324 BC (Strabo, *Geogr.* 15.1.28; Plutarch, *Alexander* 66.3). The third title, *praefectus classis*, equates the Ptolemaic office with a Roman one that seems to have developed by the time of the Second Punic

---

1. W. W. Tarn, "Two Notes on Ptolemaic History," *JHS* 53 (1933): 57–68.

War in the late third century BC (Pliny, *Nat. hist.* 6.183; Livy, *Ab urbe condita* 26.48.14).

The two Greek terms may also represent different stages in Timosthenes's career: it is possible that he was first a squadron leader (*archikybernetes*) and then overall naval commander (*nauarchos*). Several holders of the latter office are known during the reigns of Ptolemy I and II, and their probable sequence suggests that Timosthenes's tenure was near the end of the reign of Ptolemy II, perhaps after 256 BC.[2] Nothing else is known about his career other than the topographical evidence implicit in the fragments of *On Harbors*, and here, as always, there is the problem of whether written works represent autopsy or received information on the part of the author, an issue as old as Homer and Herodotus. Timosthenes visited some of the places that he wrote about (see F2)—at least in the Mediterranean, Black Sea, and Aithiopia—and worked in the Ptolemaic naval headquarters in Alexandria, where he collated various reports from field personnel and turned them into *On Harbors*. Any trip to the western Mediterranean was probably before 264 BC: the outbreak of the First Punic War would have made travel by noncombatants into that region difficult, and the Ptolemies sought to maintain neutrality.[3]

Timosthenes may also have written on cults and religion (F36–7), but the evidence for this is even more limited than for *On Harbors*. Another title, *Stadiasmos*, a one-book summary cited only by two late authors (Markianos, *Epitome* 3; Stephanos of Byzantion, "Agathe."), is probably nothing more than an abridged version of *On Harbors* that merely provided distances. Vague references to Timosthenes as the author of coastal sailing manuals (*periploi*) presumably only refer to the data in *On Harbors* (Agathemeros 7; Markianos, *Epitome* 3).

Timosthenes's *On Harbors* (Περὶ Λιμένων) was ten books in length and thus would have been longer than many extant works of Greek literature; yet the mere handful of surviving fragments is demonstrative of the tremendous loss of ancient texts, especially geographical writings. It was composed around the middle of the third century BC and was available to Hellenistic geographers: Pliny was probably the last to see a copy, although it is often difficult to determine whether a later author actually had a copy of an earlier work or was merely quoting it from a secondary source. There have only been two previous editions of the fragments of *On Harbors*: that of Emil August Wagner in 1884, and the entry in the new *FGrHist* from 2013.[4] The present edition is the first in English.

2. Irwin L. Merker, "The Ptolemaic Officials and the League of the Islanders," *Historia* 19 (1970): 153–54; Tarn, "Two Notes," 67.

3. Günther Hölbl, *A History of the Ptolemaic Empire*, trans. Tina Saavedra (London: Routledge, 2001), 54; P. M. Fraser, *Ptolemaic Alexandria* (Oxford: Clarendon, 1972), 1:152.

4. Emil August Wagner, *Die Erdbeschreibung des Timosthenes von Rhodes* (Leipzig: Frankenstein & Wagner, 1884); *FGrHist* #2051 (ed. Doris Meyer); see also Friedrich Gisinger, "Timosthenes von Rhodos [#3]," *RE*, second series, 6 (1937): 1310–22.

The extant fragments are preserved by nine authors, from the writer of the geographical poem known conventionally as Pseudo-Skymnos, dating to the second half of the second century BC, to Stephanos of Byzantion in the sixth century AD. Strabo and Pliny were responsible for nearly half the fragments, as might be expected. Unusual, perhaps, is the large number preserved in scholia, nine in all; these, as well as the five in the *Ethnika* of Stephanos, suggest that *On Harbors* was particularly prominent in late antiquity, perhaps in the one-book summary known as *Stadiasmos* that was known to Markianos and Stephanos. In addition to the extant authors who preserved fragments, the treatise was known to Eratosthenes, Hipparchos, Polybios, Poseidonios, Alexandros Polyhistor, and Didymos Chalkenteros. Awareness of the work by these authors, as well as the extant citations by Pseudo-Skymnos and Strabo, demonstrates that the work was important during the last two centuries BC and was perhaps the definitive treatment of its topic as long as the Ptolemaic empire survived. It also continued to be available in the Roman period—whether derivatively or directly—as shown through citations by Pliny, Agathemeros, and Ptolemy the geographer.

Because the fragments are so scant, it is difficult to determine the overall scope of *On Harbors*. The fragments are almost entirely limited to navigational and topographical data, with little preserved that provides any history or ethnology. If Timosthenes included such material, it was probably excised from the *Stadiasmos*, which may have been the primary means of access to the work in later times. Regardless of what the complete *On Harbors* contained, it was considered a reliable geographical treatise into the Roman period, with some allowance made for what was seen as inadequate knowledge of the western Mediterranean, an easy charge given the vastly superior understanding of the region by late Hellenistic times.

On Harbors was primarily a navigational guide for Ptolemaic seamen. This can best be seen in Timosthenes's cataloguing of celestial phenomena important for sailing (F9), as well as his innovative work in determining the winds and creating what eventually came to be called a "wind rose" (F6–8). Understanding the winds was an essential tool for sailors, and failure to do so could cause great difficulty or unusual results. Odysseus knew this, as did, perhaps less mythologically, Kolaios of Samos, who around 630 BC discovered the mineral rich regions of southwestern Iberia allegedly because of unexpected winds (Herodotus, *Hist.* 4.152).

Needless to say, sailors knew about the winds long before anything was written down, and the earliest Greek literature shows solid awareness of the prevailing winds in the Mediterranean. Homer knew of four winds (*notos, boreas, euros, zephyros*), roughly corresponding to north, south, east, and west (*Od.* 5.331–32). These winds became redefined and elaborated upon in later times.

Since *On Harbors* was a practical manual, Timosthenes probably had little interest in wind theory—something first developed by the Ionian monists Anaximandros (F32 [Graham]) and Anaximenes (F28 [Graham])—but he wanted to provide as precise a definition of the varying winds as possible.

Most of the remaining fragments of *On Harbors* are topographical. It is hardly possible to arrange them in any certain order. The only book numbers preserved are from Book 5, which included a discussion of Nikaia in central Greece (F18), and Book 6, which contained material on the island of Salamis (F19). Thus the geographical order of this portion of the treatise was south through the Greek peninsula, and since Nikaia and Salamis are not far apart, there is some hint of the depth of detail in the work.

On Harbors seems to have opened with a discussion of the extent of the inhabited world and its continents (F10), with the unusual worldview that there were four continents (separating Egypt from Libya), instead of the two or three assumed in Greek geographical theory since the fifth century BC.[5] The primary focus of the treatise is the Mediterranean and Black Sea, yet several fragments are from other regions that would also be valuable to Ptolemaic navigators. A preserved distance from Kanobos (near Alexandria) to Sebennytos (inland on the Bousiritic mouth of the Nile) may be the sole surviving remnant of a network of sailing routes through the waterine maze of the Nile Delta (F11). Additionally, there is a report of the journey from Syene to Meroë and beyond on the upper Nile (F12). Timosthenes may have made this voyage himself, but he probably also collated reports sent from this region by the several explorers commissioned by Ptolemy II to go well upstream into Aithiopia. *On Harbors* further provided detailed information on the shape, size, and nature of the Red Sea (F13–14).

The remaining fragments are largely topographical data with little context and gleaned from the treatise by later sources. Sailing distances are frequently reported (F15–16, 29–32). Presumably these are remnants of a grid of shipping lanes that covered almost the entire Mediterranean and Black Sea. The emphasis on distances in the extant fragments is probably due to the survival of the *Stadiasmos* after the complete treatise had been lost, but the recording of distances was probably a central focus of the original work. Yet there are many extant details that show the utility of *On Harbors* to mariners, who could learn that near Nikaia in central Greece there was an anchorage for a large ship (F18), or that at Artake near Kyzikos in the Propontis there was "a deep harbor for eight ships" (F23).

Occasionally coastal mountains and mouths of rivers were recorded (F20–1, 23, 28), also useful reference points for sailors. Ethnic coloration is limited, but

---

5. Duane W. Roller, *Ancient Geography: The Discovery of the World in Ancient Greece and Rome* (London: Tauris, 2015), 50–51.

some of the preserved details could have value to seamen and fall into the category of coastal landmarks (F24, 28, 34). But the only hint of a more complete ethnology is in the region of the eastern Black Sea, where Timosthenes showed interest in the local mythology relating to Phrixos and the Argonauts (F25) and the great market and trade center of Dioskourias (modern Sukhumi in Georgia), perhaps important data for the Ptolemaic trade network (F27).

On Harbors included a survey of the western Mediterranean. One can be dismissive of the charges made by later authors, especially Polybios and Strabo, that Timosthenes was ignorant of this region, a belief resulting from the greater knowledge of the west in late Hellenistic and Roman times. To be sure, it seems that Ptolemaic interests did not include the Adriatic or, wisely, the Etruscan and Carthaginian territories (F31). Yet Timosthenes provided distances west from Massalia along the Ligurian and Iberian coast as far as the Pillars of Herakles (F32–34).

Two of the extant fragments cannot plausibly be connected to *On Harbors* or any topographical writings (F36–37). One concerns a name of Persephone, Daira, and the other the cultic music at Delphi. Like so many Hellenistic intellectuals, Timosthenes's talents were not limited to one area of expertise, and these fragments may come from a cultic work titled *Exegetikon*.

The Fragments of Timosthenes's Writings

Following is a new collation of the extant fragments based on those in the Wagner (W) and *BNJ* editions, but renumbered, in order better to reflect the geographical orientation of *On Harbors*. Each fragment has an attached commentary.

Life, Works, and Reputation

1. Markianos of Herakleia, Epitome of the Periplous of Menippos *2 (W1, BNJ T1).*

The following seem to be the ones who have examined these things rationally: Timosthenes the Rhodian, who was chief pilot for the second Ptolemy, and after him Eratosthenes, whom those in charge of the Mouseion called Beta. Along with them there were Pytheas the Massaliote, Isidoros the Charakene, the pilot Sosandros, who wrote about India, and Simmias, who put together a *periplous* of the inhabited world. Moreover, there are Apelles the Kyrenaian, Euthymenes the Massaliote, Phileas the Athenian, Androsthenes the Thasian, Kleon the Sicilian, Eudoxos the Rhodian, and Hanno the Carthaginian. Some of them wrote about part of the Internal Sea, and others about all of it, and still others composed *periploi* about the External Sea.

Commentary: Markianos (Marcianus) of Herakleia, who lived at some uncertain date after AD 250, produced an edition of the three-book *Periplous* of Menippos of Pergamon. Although Markianos made corrections and additions to the earlier work, these seem to have been minor, and the two places where his epitome cited Timosthenes (F1–2) can be assumed to have been from Menippos with perhaps some adjustments in tone.[6] Menippos was active in the first century BC. He was mentioned, along with his geographical writings, by Krinagoras of Mytilene, the Augustan court poet, in the context of a trip to Italy by the latter, probably in 26/25 BC (*Greek Anthology* 9.559); thus, Menippos was a contemporary of Strabo.

His *periplous* seems to have been similar to Timosthenes's *On Harbors*, emphasizing navigational information, which may be why he ranked Timosthenes first among his geographical precedessors, certainly a unique point of view. The list of writers in F1 is in no obvious order—certainly not chronological—and may indicate their utility to Menippos. Nevertheless, it is unsually thorough: Sosandros and Apelles are cited nowhere else in extant literature. F1 is the only source to report that Timosthenes was from Rhodes, and one of the few (see also F12, 36) to provide his title of ἀρχικυβερνήτης and to note his association with Ptolemy II. If Menippos relied heavily on the earlier scholar, these details may carry particular weight.

2. Markianos of Herakleia, Epitome of the Periplous of Menippos *3 (W3, BNJ T2, 4, 6).*

Timosthenes, when he wrote his books *On Harbors*, did not examine accurately all the peoples who lived around our sea, since most portions of the sea were still unknown (the Romans had not yet conquered it through warfare). In regard to Europe, he neglected the Tyrrhenian Sea, having only incompletely sailed around it, and his knowledge of places adjacent to the Strait of Herakles—on either the Internal or External Sea—was lacking. Moreover, concerning Libya, he was completely ignorant of the places extending from Karchedon through the Herakleian Strait and into the External Sea. There is an epitome in one book of his ten books, and there is another one, called the *Stadiasmos*, which is a summary of what he wrote. In all of these there is nothing reported of accomplishment or distinction. Eratosthenes the Kyrenaian rewrote the book of Timosthenes—I do not know for what reason—adding some small things, so that he would not differ greatly in his introduction with what had been previously recorded, but in order to set forth in his own words what had been said by others.

---

6. Aubrey Diller, *The Tradition of the Minor Greek Geographers* (Lancaster, PA: Lancaster Press, 1952), 147–50.

Commentary: For Markianos and Menippos, see F1. Menippos, like other writers of his era (see also F31, 33, 35), criticized Timosthenes for his perceived ignorance of the western Mediterranean, but also reasonably noted that it was less known before the Roman period. Yet some of Menippos's comments were contradicted elsewhere: Timosthenes did know about the western coast of North Africa and the Herakleian Strait (F32–33). This fragment is unique in stating that Timosthenes actually did field research, however incomplete. Only Menippos and his contemporary Strabo (F36) provide the ten-book length of *On Harbors*, and if all the comments about Timosthenes in F2 derive from Menippos and are not by his epitomizer Markianos, it seems that the summary known as the *Stadiasmos* existed by the first century BC.

The statement that Eratosthenes's *Geography* was merely a reworking of *On Harbors* is bizarre, to say the least, and could hardly have been legitimately said by anyone who had read Eratosthenes's treatise, who nevertheless did have a high opinion of the earlier scholar (F3). If it is not merely an attempt to impress readers with the importance of *On Harbors*—perhaps more to Menippos's liking than the later work—the comment may reflect the hand of Markianos, who may not have had either available. Nevertheless, Markianos's account has a rather contradictory tone: it was said that there was nothing of note in Timosthenes's treatise, but it was so significant that Eratosthenes, the inventor of the discipline of geography, heavily derived his own *Geography* from it.

3. Strabo, Geography *2.1.40 (W2, BNJ T7).*

But the errors made about them [the promontories of Europe] by Eratosthenes [*Geography,* F134] are so numerous, as well as those by Timosthenes, the writer of *On Harbors*, whom he praises more than the others (although he refutes him, disagreeing on most things).

Commentary: Like Menippos (F1–2), Strabo expressed his concern about Timosthenes's apparent ignorance of the western Mediterranean. The focus of his complaint is not certain: he has defined southern Europe in terms of promontories (the Greek peninsula, Italy, and Iberia), but he probably viewed Timosthenes as deficient only in his knowledge of the last two. Also like Menippos, Strabo acknowledged the debt of Eratosthenes to Timosthenes but in a more nuanced way.

4. Strabo, Geography *2.1.41 (W4, BNJ T5, 8).*

In this section he [Eratosthenes] and Timosthenes provide just occasion [for criticism], so I will abandon any joint examination, as what Hipparchos [*Against the Geography of Eratosthenes,* F34] has said about this is sufficient.

Commentary: F4 presents Strabo's concluding remarks about the failure of earlier geographers to understand the western Mediterranean. His comments are directed mostly against Eratosthenes and Hipparchos (the second century BC author of *Against the Geography of Eratosthenes*), but these comments provide the only evidence that Hipparchos was familiar with Timosthenes's work.

5. Pseudo-Skymnos 109–119 (BNJ T9).

I will now come to the beginning of my treatise, setting forth the writers whom I have used to move my historical account toward reliability, especially the one who has written most accurately about geography—both the latitudes and figures—Eratosthenes, as well as Ephoros, and the one who spoke in five books about foundations (the Chalkidean Dionysios), and also the writers Demetrios the Kallatianian, Kleon of Sicily, and Timosthenes . . .

Commentary: The work by an unknown author who goes under the modern appellation of Pseudo-Skymnos is a geographical poem of over seven hundred lines.[7] The name of its author is of less importance than its date, which can be determined with unusual precision, since it is dedicated to a King Nikomedes, and internal evidence within the poem demonstrates that this is the third king of that name, who ruled Bithynia ca. 127–94 BC. Thus the poem is the earliest extant mention of Timosthenes. The text is poorly preserved at this point and essentially breaks off for several lines after the citation of Timosthenes, so the exact structure of the list cannot be determined. If the author had anything specific to say about Timosthenes, it is lost. The five authors cited before to him range from those of great importance in the history of geography (Eratosthenes and Ephoros) to some who are obscurities to the modern reader (Dionysios and Kleon). What the author gleaned from *On Harbors* remains uncertain, since his poem is more ethnographically oriented than Timosthenes's treatise.

Winds and Navigation

6. Agathemeros 7 (W6b, BNJ F3).

Timosthenes, who wrote *periploi*, says that there are twelve [winds], adding the *boreas* between the *aparktias* and *kaikias*, the *phoinix* (or *euronotos*) between the *euros* and *notos*, the *leukonotos* or *libonotos* between the *notos* and *lips*, and the *thraskias* (called *kirkios* by the locals) between the *aparktias* and *argestes*. The following peoples live at the boundaries of the earth: toward the

---

7. Didier Marcotte, ed., *Géographes grecs*, vol. 1, *Introduction générale; Ps.-Scymnos, Circuit de la Terre* (Paris: Les Belles Lettres, 2000), 35–46.

apeliotes are the Baktrianians, toward the *euros* the Indians, toward the *phoinix* the Erythra Sea and Aithiopia, toward the *notos* Aithiopia beyond Egypt, toward the *leukonotos* the Garamantians beyond the Syrtes, toward the *lips* the Western Aithiopians beyond the Maurians, toward the *zephyros* the Pillars and the beginning of Libya and Europe, toward the *argestes* Iberia (now called Hispania), toward the *thraskias* <the Kelts and those bordering on them, toward the *aparktias*> the Skythians beyond Thrace, toward the *boreas* the Pontos and the Maiotic Sarmatians, and toward the *kaikias* the Caspian Sea and the Sakians.

Commentary: Agathemeros, the author of a brief geographical treatise, is hardly known. His work was written after the time of Menippos of Pergamon in the Augustan period, and no later than the second century AD.[8] The author had a particular interest in the winds, since he recorded in detail Timosthenes's system, the only extant author to do so. After laying out eight winds, essentially the scheme of Aristotle (*Meteorologika* 2.6), he then credited Timosthenes with identifying twelve, which are outlined in a clockwise fashion from the north. Although there was never agreement in antiquity about the number of winds—the extant Tower of the Winds in Athens, dating to the first century BC, includes eight winds, and Vitruvius, writing contemporary to the construction of the tower, reported twenty (*On Architecture* 1.6.10)—Timosthenes's system became commonly used.[9] He also added some new wind names: the *libonotos* ("Libyan-south") or *leukonotos* ("white-south") for the southwest wind, and the *kirkios*, a local variant for the northwest wind, generally called the *thraskias* ("Thracian").[10] Timosthenes also provided data on the length of time that seasonal winds blew, something of importance to sailors: the etesian winds (from the northwest in summer) lasted for fifty days from mid-July into September (F8).

Unlike Aristotle, Timosthenes gave the winds an ethnic orientation, perhaps showing awareness of Ephoros's famous statement describing the perimeter of the inhabited world in terms of ethnic groups (Ephoros, F30a = Strabo, *Geogr.* 1.2.28). The list presented by Agathemeros has been slightly altered, since "Hispania" is a Latin term and is not documented before the first century BC (Cicero, *On the Imperium of Pompeius* 30; Caesar, *Gallic War* 1.1). Clearly any wind system depended on a point of reference, and this may be why Timosthenes added the ethnic component. His system would be applicable for the two places associated with his career, Rhodes and Alexandria, which lie only 375 miles apart. If the system were intended for Ptolemaic use, Alexandria would be the

8. Aubrey Diller, "Agathemerus, *Sketch of Geography*," *GRBS* 16 (1975): 72.

9. For a diagram, see Ian G. Kidd, *Posidonius*, vol. 2, *The Commentary* (Cambridge: Cambridge University Press, 1988), 520.

10. For the winds themselves, and the meaning of their names, see Paul Moraux, "Anecdota Graeca Minora II: Über die Winde," *ZPE* 41 (1981): 43–58.

better central point, but some of its elements (such as the *zephyros* blowing from the Pillars of Herakles and the *leukonotos* from the Syrtes) are more appropriate for Rhodes, which had a long maritime history and probably had a vast amount of data available to Timosthenes. The orientation of the wind system on the island may be the reason that Eratosthenes chose it as the crossing point of his prime meridian and prime parallel. He also attempted to refine Timosthenes's system in order to make it universally applicable.[11]

This discussion of the winds and their relationships to ethnic groups may have been part of the introduction to *On Harbors*. Except for the insertion of Hispania, it represents a worldview of the generations immediately after Alexander the Great, with the extremities of the inhabited portions extending from India, Baktria, and the Caspian Sea across Africa to the Pillars of Herakles. Notably absent, and supporting the criticisms of Strabo and Menippos (F2–4), is any mention of the peoples of western Europe except an uncertain reference to the Kelts, who have been inserted into a deficient portion of the text and may not be the ethnym that Timosthenes used.

7. Strabo, Geography *1.2.21 [W6a, BNJ F6].*

But Poseidonios [F137a] says that none of the authorities (such as Aristotle, Timosthenes, or Bion the Astronomer) have reported this about the winds.

Commentary: Strabo's context is that some authorities recognized only two major winds—*boreas* and *notos*—and that the others were merely variants. He may have believed that this was the view of Homer, which in his mind gave it authority, but he also noted that Poseidonios soundly rejected this idea (Homer, *Il.* 9.5; 11.306; 21.334; Poseidonios F137a). The argument is tangled into Strabo's incessant defense of Homer and adds nothing new about Timosthenes, but the testimonium is important because it is the only evidence that Poseidonios had read *On Harbors*.

8. Scholia to Apollonios, Argonautica *2.526 (W7, BNJ F5).*

Some say that the etesian winds last for forty days: others, such as Timosthenes, say fifty. They begin to blow when the sun leaves Karkinos [Cancer] and blow throughout Leo, abating two-thirds of the way through Parthenos [Virgo].

Commentary: The scholia to Apollonios's *Argonautica* are remarkable sources for fragments of *On Harbors*, providing six citations. They date, in part,

---

11. Eratosthenes, *Geography* F11 (= Strabo, *Geogr.* 1.2.20), F56 (= Strabo, *Geogr.* 2.1.33); O. A. W. Dilke, *Greek and Roman Maps* (Ithaca: Cornell University Press, 1985), 31–32; Phillip Harding, *Didymos on Demosthenes* (Oxford: Clarendon, 2006), 30; Kidd, *Posidonius*, 521–22.

from as early as the first century BC,[12] so some of the material may come from a period when *On Harbors* was still well known. The etesian winds, first mentioned by Herodotus (*Hist.* 2.20; 6.140), are the northwesterlies that blew every year (*etos*) during the Mediterranean summer: Timosthenes defined their duration precisely, something that mariners needed to know (see also F6).

9. *Scholia to Apollonios,* Argonautica *2.517 (W42,* BNJ *F31).*

Sirius: some say that the star is the dog of Orion, others that of Erigone, or that of Alkyoneus, or of Isis, or of Kephalos, and still others—according to Timosthenes—that it is a proper name.

Commentary: For the scholia to Apollonios see F8. Understanding the stars would be essential to navigators, and one can assume that *On Harbors* contained extensive data on celestial phenomena, of which this is the only remnant. The heliacal rising of Sirius had been an important element of the annual cycle since before the Greeks, and its association with the heat of summer was also ancient (Hesiod, *Works and Days* 586–588). "Sirius" (Greek "Seirios") is conventionally derived from Greek σιγάω, "to scorch," but it is by no means certain whether the star name or the verb came earlier, as the latter first appears in Aratos's *Phainomena* (331), contemporary with *On Harbors.* Timosthenes's suggestion that it was originally a proper name may be accurate; the word is possibly pre-Greek. His demonstrated linguistic interest is a rare example of a comment in *On Harbors* not directly related to seamanship.

The Continents

10. *Bern Commentary to Lucan (ed. Usener) 9.411 (W8,* BNJ *F1).*

Libya is the third part of the world, if one believes the reports: some, such as Varro [*De lingua latina* 5.31; *De re rustica* 1.2.3], divide the world into two parts (Asia and Europe), others, such as Alexander, into three (Asia, Europe, and Africa), and others (such as Timosthenes) into four, adding Egypt.

Commentary: The Bern scholia to Lucan (*Commenta Bernensia*) dates from the tenth century AD.[13] It is the only source for the unusual statement that Timosthenes believed that there were four continents (instead of the usual three). Continental theory seems to have been originated by Hekataios of Miletos around

---

12. Eleanor Dickey, *Ancient Greek Scholarship: A Guide to Finding, Reading, and Understanding Scholia, Commentaries, Lexica, and Grammatical Treatises, from Their Beginnings to the Byzantine Period* (Oxford: Oxford University Press, 2007), 62–63.

13. Hermann Usener, *Scholia in Lucani Bellum Civile* (Leipzig: Teubner, 1889).

500 BC and developed by Herodotus (*Hist.* 4.45); originally there were two (Europe and Asia), but Libya had been added by the fifth century BC. The division between Libya and Asia was always a problem,[14] and this may have led Timosthenes to suggest that Egypt was a separate continent (perhaps an element of Ptolemaic ideology), but it is quite possible that the commentator misunderstood what Timosthenes said, since it is clear that he misunderstood Varro, who was not writing about continents but about Eratosthenes's division of the inhabited world into a northern and southern part, defined by the east-west orientation of the Mediterranean and the Tauros Mountains (Eratosthenes, *Geography* F47).

Egypt and Aithiopia

11. Ptolemy, Geographical Guide *1.15.4 (W16, BNJ F20).*

From Kanobos to Sebennytos was established by Timosthenes at 290 [stadia], but this distance should be greater.

Commentary: The *Geographical Guide* of Ptolemy was written in the mid-second century AD, and was based on a slightly earlier work by an elusive Marinos of Tyre. Ptolemy, who quoted Timosthenes twice (see also F29) may only have had access to (or interest in) the condensed *Stadiasmos*, the table of distances. Kanobos (at modern Abukir) was just northeast of Alexandria near the Kanobic mouth of the Nile. Sebennytos was at modern Sammanud, about fifty miles up the Bousiritic mouth. The two places are about eighty direct miles apart, but any route between them would not be overland but circuitously by water. Yet the distance of 290 stadia (perhaps thirty-five miles) is far too short by any means, as Ptolemy knew, and there has been some problem with the transmission of the data.

12. Pliny, Natural History *6.183 (W37, BNJ F19).*

Similarly there have been various reports about the measurement [of Aithiopia], first by Dalion, who sailed far beyond Meroë, and later by Aristokreon, Bion, and Basilis, as well as by the younger Simonides, who stayed for five years in Meroë while writing on Aithiopia. Moreover, Timosthenes, the commander of the fleet of Philadelphos, has reported that the journey from Syene to Meroë takes sixty days, without giving the daily measurement.

Commentary: Pliny, whose *Natural History* was completed by the AD 70s, may have been the last writer to make direct use of a copy of *On Harbors*. He listed five explorers of the upper Nile, who went to the Aithiopian capital of Meroë and beyond, perhaps to somewhere in the vicinity of modern Khartoum. One of them,

14. Roller, *Ancient Geography*, 50–51.

Simonides, lived at Meroë for five years, presumably as Ptolemaic agent. No date is given for any of the explorers, but it can be presumed they were sent by the Ptolemaic government during the third century BC, and some would have been known to Timosthenes. As a separate item Pliny noted that Timosthenes recorded the travel time from Syene (at the First Cataract) to Meroë (sixty days). This is close to Herodotus's report of about fifty-six days (*Hist.* 2.29), and Timosthenes may merely have been transmitting his data, although one would expect that he had more recent information. The structure of Pliny's account, which separates Timosthenes from the five explorers, suggests that the latter did not travel to the upper Nile himself but merely collated earlier reports, but this is far from certain.

The Red Sea and the Eastern Ocean

13. Pliny, Natural History *6.163 (W9, BNJ F21).*

We will now follow the remainder of the coast opposite Arabia. Timosthenes figured the length of the entire gulf [of the Red Sea] at four [?] days' sail and two in width, and seven and a half miles at its narrowest.

Commentary: Knowledge of the Red Sea was a Ptolemaic priority, since it was the route to East Africa, the Arabian aromatic region, and India. More importantly, however, was the need for elephants, an essential part of Hellenistic warfare, which were found along the southern coasts of the African side of the sea (Strabo, *Geogr.* 16.4.7; 17.1.5). The Red Sea had first been explored by Anaxikrates, commissioned by Alexander the Great in his last days (Eratosthenes, *Geography* F95 = Strabo, *Geogr.* 16.2.4). He recorded its length at 14,000 stadia (about 1,750 miles), somewhat too long but perhaps not reflecting a direct route. Yet Timosthenes's four sailing days is a major error, probably due to erroneous copying of the source or textual transmission: 14,000 stadia would take weeks, not days, and as presented the Red Sea is only twice as long as it is wide. Timosthenes's original number may have been fourteen.[15] But clearly he was more interested in travel times for Ptolemaic shipping than actual distances. F13 and F14 are the only evidence that *On Harbors* included data on sea lanes outside the Mediterranean and Black Sea.

14. Pliny, Natural History *6.198 (W10, BNJ F22).*

Ephoros [F172], Eudoxos [F369], and Timosthenes record that there are many islands throughout the Eastern Sea.

---

15. For a list of known sailing times, see Lionel Casson, *Ships and Seamanship in the Ancient World* (Princeton: Princeton University Press, 1971), 282–96.

Commentary: "Eastern Sea" (*Eoo mari*) seems to refer to the Red Sea (the *Mare Ruber* of Pliny, cited in chapters 196 and 199). Yet it is a peculiar term, and in fact most of the discussion of chapter 198 is about the west coast of Africa, far beyond Timosthenes's area of interest. But the fact remains that the Red Sea was an area of vital Ptolemaic concern, and the confusion is probably due to Pliny. There are hundreds of islands in the Red Sea, some of which were navigational hazards. Other islands posed different kinds of threats, such as the ominously named Ophiodes ("Snaky"; see Strabo, *Geogr.* 16.4.6–7). There is nothing particularly profound about Timosthenes's statement, and it was probably an introduction to a detailed discussion of the islands in the Red Sea, particularly those that affected shipping.

The Eastern Mediterranean

15. Pliny, Natural History *5.47 (W30, BNJ F2).*

Asia adjoins [Africa], whose extent—from the Canopic mouth to the mouth of the Pontos—Timosthenes records as 2,638 miles.

Commentary: The distance from Alexandria (actually the Canopic mouth of the Nile just to its northeast) to the mouth of the Pontos (Black Sea) at the northeast end of the Thracian Bosporos would have been an important datum for Ptolemaic seamen. The number (converted from stadia to Roman miles, equal to 2,427 statute miles) is absurd; it is actually less than 1,000 miles, and either it is a large error or, more likely, an actual sailing route winding through the Aegean islands.

16. Pliny, Natural History *5.129 (W26, BNJ F23).*

The circumference [of Cyprus] is reported by Timosthenes as 427½ miles and by Isidoros [F13] as 375. Its length between the two promontories of Klidai and Akamas (which is toward the sunset) is 162½ according to Artemidoros [F117] and 200 according to Timosthenes.

Commentary: A common technique used by Pliny in the geographical portions of his *Natural History* was to compare distances from several sources, usually converted from the original to Roman miles. Both of the authors compared to Timosthenes are later than him: Artemidoros lived around 100 BC and wrote a lengthy geographical treatise, and Isidoros was active in the Augustan period and was the author of the extant *Parthian Stations* and other geographical works. Distances resulting from circumnavigations are always problematic, since the exact route cannot be known, but both figures for the circumference of Cyprus are below the modern one (equivalent to 438 Roman miles), although

Timosthenes's estimation is quite close. The length of Cyprus is calculated today at the equivalent of 153 Roman miles, much lower than the ancient figures, but again the ancient text does not record the route taken.

The Aegean

17. Scholia to Apollonios, Argonautica *4.1712 (W24,* BNJ *F26).*

The island of Hippouris is near Thera: Timosthenes mentions it.
  Commentary: For the Scholia to Apollonios, see F8. Hippouris (modern Phtini) is one of the most obscure of Greek islands, about fifteen miles east-southeast of Thera and just south of Anaphe. There are actually two tiny islets, virtually unknown in ancient history. The island of Hippurus mentioned by Pomponius Mela (*Chor.* 2.111) may be another island, farther east. Why Timosthenes should have mentioned Hippouris remains uncertain, but it is interesting that the two certain mention of this island in Greek literature, his and that of Apollonios (*Argonautica* 4.1712), are virtually contemporary and were both reported by sources connected with Rhodes. The island may have played some unknown role in Ptolemaic or Rhodian seamanship, perhaps as a marker on the sailing route from Rhodes to the western Mediterranean.

18. Didymos Chalkenteros on Demosthenes, Response to Philip's Letter *(*BNJ *F12).*

Nikaia is a city on the sea, twenty stadia distant from Thermopylai, about which Timosthenes in Book 5 of his *On Harbors* writes the following: "Going from Thermopylai by ship it is about twenty stadia to Nikaia, and on foot it is about fifty. Approximately five stadia distant is a sandy promontory and in another four stadia there is an anchorage for a large ship."
  Commentary: Didymos "Chalkenteros" of Alexandria was a major scholar of the first century BC, noted for his analyses of earlier works. In Demosthenes's *Response to Philip's Letter,* the orator noted that Philip II of Macedonia kept a garrison at Nikaia (the context is 340 BC), on the Lokrian-Phokian border just east of Thermopylai (at modern Roumelio or Platanakos; Demosthenes, *Response to Philip's Letter* (11) 4; see also Strabo, *Geogr.* 9.4.4). Timosthenes's interest was not so much the locality itself but a nearby anchorage that could be used by large ships. F18 is also important because it is one of the few fragments that provide a book number for a passage in *On Harbors* (see also F19), which suggests that Central Greece was about halfway through the treatise. It also has one of only two direct quotations from the work (see also F23), and thus gives a sense of the tone of the treatise.

19. Scholia to Aeschylus, Persians *303 (W40,* BNJ *F13).*

Sileniai is a beach on Salamis, near the so-called Tropaion Promontory, according to Timoxenos [presumably Timosthenes] in the sixth book of his *On Harbors*.

Commentary: Some of the scholia to Aeschylus are as early as the Hellenistic period.[16] Sileniai (or Selenia), on Salamis, figured in the battle of 480 BC. It was located on the eastern shore of the island near the victory trophy for the battle, which probably stood on Kynosoura, the long eastern peninsula (Aeschylus, *Persians* 303; Pausanias, *Descr.* 1.23.6).[17] Having discussed Central Greece in Book 5 of *On Harbors* (F18), by the sixth book the account has moved to the south.

Northwest Anatolia

20. Stephanos of Byzantion, "Apia" (W33, BNJ *F7).*

[Demetrios of Magnesia] says that the Apidanos River is in the Troad, emptying into the western sea, as does Timosthenes.

Commentary: The *Ethnika* of Stephanos of Byzantion, a grammatical work of probably the sixth century AD, contains a lot of geographical information, including five citations from *On Harbors* (see also F21, 23, 31, 35). Demetrios of Magnesia was a grammarian of the first century BC who wrote on homonyms and dedicated a work to Cicero's friend Atticus (Dionysios of Halikarnassos, *On Deinarchos* 1; Cicero, *Letters to Atticus* #162). He may have been Stephanos's source for Timosthenes's item. The Apidanos River in the Troad is otherwise unknown, but the mouths of rivers would have been important to navigators.

21. Stephanos of Byzantion, "Alexandreai" (W34, BNJ *F6).*

... there is a place on Ida in the Troad called Alexandreia, where Paris made the judgement of the goddesses, according to Timosthenes.

Commentary: For Stephanos of Byzantion, see F20. Alexandreia, or Alexandria, is presumably Alexandria Troas, at modern Dalyanköy on the coast a few miles southwest of ancient Troy. The statement is vague and inaccurate: Alexandria is not on Mt. Ida, which lies to the east-southeast, on which the

---

16. Dickey, *Ancient Greek Scholarship*, 35–38.

17. A. J. Podlecki, *The Life of Themistocles: A Critical Survey of the Literary and Archaeological Evidence* (Montreal: McGill-Queen's University Press, 1975), 122–23.

judgment of Paris took place. The fragment seems to be a poor conflation of several pieces of data; if Timosthenes reported on Paris, it is a rare example of mythological interest on his part (see also F25).

22. Strabo, Geography *13.2.5 (W22, BNJ F24).*

In the strait between Asia and Lesbos there are about twenty islets, but Timosthenes says forty. They are called the Hekatonnesoi.

Commentary: There are many islands in the channel between Lesbos and the Anatolian mainland. The Hekatonnesoi ("Hundred Islands") are hard to identify, as demonstrated by the uncertainty as to their numbers (100, 40, or 20). The difficulty may be that the name is not numeric but descriptive, after a local cult, Apollo Hekatos (Strabo, *Geogr.* 13.2.5). Nevertheless, an area with so many tiny islands would be a navigational hazard.

23. Stephanos of Byzantion, "Artake" (W31, BNJ F8).

Artake, a Phrygian city and Milesian settlement. Demetrios [of Magnesia] says that it is a small island, and Timosthenes says "Artake is, first, a mountain in the Kyzikene, and also a small island, a stadion distant from the mainland. It has a deep harbor for eight ships, below the headland that the mountain creates near the beach."

Commentary: For Stephanos of Byzantion and Demetrios see F21. Artake is a toponym in the vicinity of Kyzikos on the Propontis. The mountain is probably the one that forms the Kyzikene peninsula, usually called Arktonnesos (Bear Island), probably a name related to Artake. The city, at modern Erdek, was on the southwest side of the peninsula, and the island, modern Tavşan Adası, is just offshore (Strabo, *Geogr.* 12.8.11). Timosthenes's interest, as usual, was navigational, noting the harbor and its capacity. This is the second of two surviving direct quotations from *On Harbors* (see also F18).

The Black Sea

24. Harpokration, "To Hieron" (W27, BNJ F90).

To Hieron: a sanctuary of the Twelve Gods on the Bosporos, according to Timosthenes in his *On Harbors.*

Commentary: Harpokration of Alexandria was a lexicographer of the second century AD. Hieron lies on the southern shore of the Thracian Bosporos about five miles from the mouth of the Black Sea, at modern Anadolu Kavak. It was a sanctuary founded by the people of Chalkedon (at the other end of the

Bosporos). The site may have been prominently visible from the sea, thus serving as a navigational point. F25 may have followed immediately in *On Harbors*.

25. Scholia to Apollonios, Argonautica *2.532 (W28,* BNJ *F10).*

Timosthenes says that the children of Phrixos established altars of the twelve
gods, and the Argonauts one to Poseidon.

Commentary: For the Scholia to Apollonios, see F8. F25 is probably part of
the same discussion as F24 and F26, about the route through the Thracian Bosporos and, in this case, regarding the sanctuary of the Twelve Gods at Hieron.
The extant fragments imply that Timosthenes's interest in mythology was limited (see also F21, F36), but if the altars were conspicuous from the sea, he may
have thought it suitable to comment. According to Apollonios, the altar was
actually established by the Argonauts (*Argon.* 2.531–32), but they and the sons
of Phrixos are connected, as one of Phrixos's sons, Argos, built the *Argo* and
joined the Argonauts' expedition. The altars were probably visible to ships.

26. Scholia to Theokritos, Idyll *13.22 (W26,* BNJ *F25).*

Timosthenes says that it is twenty-five stadia from Hieron to the little Skopelodes
["Rocky"] islands and what is called the Kyanea Promontory.

Commentary: The scholia to Theokritos date to as early as the Augustan period.[18] Like F24–25, this fragment is part of Timosthenes's account of
the route through the Thracian Bosporos. The Skopelodes Islands (modern
Örektaşi) are at the outlet of the Bosporos and were also known as the Symple-
gades or Kyaneai. The names—rocky, wandering, or dark—emphasize their role
as navigational hazards; they were allegedly first encountered by the Argonauts
(Herodotus, *Hist.* 4.85; Euripides, *Medea* 1263; Apollodoros, *Bibliotheke* 1.2.10).
The Kyanea Promontory was the adjacent mainland on the north side of the
Bosporos. Safe passage past these islands was essential for those entering the
Black Sea.

27. Pliny, Natural History *6.15 (W25,* BNJ *F11).*

The peoples holding all the rest of this wild coast are the Melanchlaenians
and the Coraxians, who also possess the Colchian city of Dioskourias near the
Anthemous River, now deserted, but which was once so famous that Timosthenes reported three hundred peoples with different languages used to come
down to it, and then our business was conducted by 130 interpreters.

18. Dickey, *Ancient Greek Scholarship*, 63–65.

Commentary: The city of Dioskourias lies at modern Sukhumi in Georgia, on the eastern shore of the Black Sea. It had been founded by the fifth century BC and was the seaward end of a complex group of trade routes from the Caucasus and beyond (Pseudo-Skymnos 81; Strabo, *Geogr.* 11.2.6; Pomponius, *Chor.* 1.109–110). There is no doubt that it was a major trade rendezvous center, gathering peoples from a vast area of central Asia—their contact point with the Mediterranean world—but the three hundred languages may be an exaggeration. There is no obvious reason why two ethnic groups should have been singled out from the dozens who came to the city, although both have intriguing names: the Melanchlaenians ("Black-Robed People") and Coraxians ("Raven People"). The Anthemous River is probably the modern Gumista, west of the townsite. This is the most detailed ethnographic statement preserved from *On Harbors*.

The Ionian Islands

28. Scholia to Apollonios, Argonautica *2.297 (W39, BNJ F14).*

Ainos is a mountain on Kephallenia, where there is a sanctuary of Zeus Ainesios, which Kleon records in his *Periplous* and Timosthenes in his *Harbors*.

Commentary: For the Scholia to Apollonios see F8. The Ionian island of Kephallenia is unusually rugged, with Mt. Ainos (modern Megalovouni, renamed) rising to 5,841 feet only a few miles from the coast: its summit is visible far out to sea. The sanctuary of Zeus Ainesios, whose remains survived into early modern times, was an ancient cult, and the shrine may have been a conspicuous feature to mariners (Hesiod, F104a; Strabo, *Geogr.* 10.2.15). Kleon is possibly the writer from Syracuse, probably of the fourth century BC, who also wrote an *On Harbors*, perhaps for the Syracusan kings.[19]

Sicily and Libya

29. Ptolemy, Geographical Guide *1.15.2 (W14, BNJ F28).*

Pachynos is opposite Leptis, and Himera [is opposite] Theainai. Yet the distance from Pachynos to Himera comes to about four hundred stadia, and that from Leptis to Theainai is more than 1,500, according to what Timosthenes records.

Commentary: Pachynos is the southeastern corner of Sicily (modern Capo Passero). Himera cannot be the town of that name (modern Imera) on the north shore of the island, which is much farther from Pachynos than four hundred stadia (about fifty miles), but must be the mouth of the Himera River (modern

19. Paul T. Keyser, "Kleon of Surakousai," *EANS* 481.

Salso) on the south coast, although the published distance is rather short (it is actually about seventy-five miles). Nevertheless, this is a remnant of Timosthenes's account of the coast of Sicily. Leptis (in Libya) and Theainai (more commonly Thena, in Tunisia) are on the Mediterranean coast about two hundred miles apart, close to the 1,500 stadia reported. Although the relative placement of the Sicilian and North African toponyms is attributed to Ptolemy's source, Marinos of Tyre, they may be remnants of Timosthenes's routes across the Mediterranean.

30. Agathemeros 20 (W22–3, BNJ F27).

The perimeter of Sicily, according to Timosthenes, is 4,740 stadia, with its shape that of a scalene triangle. The crossing of the strait from Cape Peloros to Italy is twelve stadia. The side of the island from Peloros to Pachynon is 1,360 stadia, and that from Pachynos to Lilybaion, 1,600 stadia. According to Timosthenes, it is 1,700 stadia from Lilybaion to Pelorias. From Lilybaion the crossing to Aspis in Libya is near to 1,500 stadia.

Commentary: For Agathemeros, see F6. He combined two sources in reporting on the coastal dimensions of Sicily: one by Timosthenes and another by an unnamed author, and the differences in the two are reflected in the variant spellings, Peloros and Pelorias.

The circumference of Sicily was of persistent interest to Greek and Roman geographers, perhaps because it was the largest island in the Mediterranean. Moreover, its proximity to the Carthaginian area of interest meant that such data were important for military reasons. The earliest known report is by Timaios of Tauromenion, a native of Sicily, who probably wrote a generation or more before Timosthenes (Timaios, *FGrHist* #566, F164 = Diodoros of Sicily, *Hist.* 5.2.2). Timaios suggested a circumference of 4,360 stadia, slightly larger than that of Timosthenes. Later, Poseidonios recorded 4,400 stadia (Poseidonios F87a = Strabo, *Geogr.* 6.2.1). These three figures—4,740, 4,400, and 4,360 stadia, or about 575 to 600 miles—are remarkably close and compare favorably with the modern official circumference of 620 miles; presumably the ancient route cut across some bays.

The distance across the Straits of Messina is reported at about 1.5 miles: it is in fact two miles at its narrowest. The itinerary around the island begins at Peloros, or Pelorias (modern Capo Peloro at the northeast corner), and goes clockwise to Pachynon (modern Capo Passero at the southeast corner), then along the southern coast to Lilybaion (modern Marsala in the southwest), and finally back around the north to Peloros/Pelorias. Given the vagaries of an indented coast the three segments around the island tend to be quite accurate, as one would expect in this important and well-known region.

Timosthenes was probably also responsible for the distance from Lilybaion to Aspis (modern Kelibia in Tunisia), the shortest crossing of the Sicilian Strait. The figure is high—about 190 miles compared to an actual distance of around one hundred miles—but presumably reflects the sailing route. This is the narrowest part of the Mediterranean, and as such would have been an important navigational point.

The Western Mediterranean

31. Strabo, Geography *2.1.41 (W18, BNJ T5).*

Let it now be said that Timosthenes, Eratosthenes [*Geography,* F131], and those even earlier were completely ignorant of the Iberian territory and Keltika, and immensely more so about the German and Brettanikian territory, as well as the territory of the Getians and the Bastarnians. They also happened to be somewhat ignorant of the region of Italia, the Adria, the Pontos, and those portions beyond to the north.

Commentary: This is not truly a fragment from *On Harbors*, yet it demonstrates the geographical extent of the work. Since Strabo coupled his objections to Timosthenes and Eratosthenes, it is difficult to separate the comments, and he showed little willingness to understand the limited knowledge of the western Mediterranean in early Hellenistic times. Moreover, some of his remarks simply are not true: Timosthenes did discuss the Keltic and Iberian territories (F32–33), as Strabo himself knew. Strabo could be excessively polemic in outlining the failures of his predecessors, and he seems to have ignored the fact that since *On Harbors* was a navigational guide, it would hardly have considered the peoples of interior Europe such as the Germans, Getians, and Bastarnians, all of whom lived along the Danube. In addition, the Germans were probably not known to the Mediterranean world in Timosthenes's day. The Etruscan territory and the Adriatic were of no interest to the Ptolemies, and the Pontos was in fact covered in *On Harbors* (F24–27). Yet Timosthenes would have had no reason to mention the peoples north of the Black Sea.

32. Stephanos of Byzantion, "Agathe" (W38, BNJ F17).

Agathe, a Ligyan or Keltic city. Skymnos says in his *Europe* that it is Phokaian. Timosthenes in his *Stadiasmos* calls it Agathe Tyche.

Commentary: Agathe, or Agathe Tyche (modern Agde) was a Massalian or Phokaian foundation at the western edge of the Rhone delta, presumably on the site of an indigenous town (Strabo, *Geogr.* 4.1.5–6). In early times the

Ligyan (Ligurian) and Keltic territories were not precisely defined, but were the populations on the modern French coast. The name, if not local, referred to its good (*agathe*) location and harbor—it was a major navigational point of the region—and had evolved by the third century BC into the common Greek phrase "Good Fortune." Skymnos of Chios (not the Pseudo-Skymnos of F5) wrote a geographical treatise in the mid-second century BC.

33. Strabo, Geography *17.3.6 (WII, BNJ F29).*

Metagonion is across from New Karchedon on the far side, but Timosthenes incorrectly says that it is across from Massalia. The crossing from New Karchedon is 3,000 stadia, and the voyage along the coast to Massalia is more than 6,000.

Commentary: Metagonion is usually identified with Cap Bougaroun in Algeria, one of the northernmost points along this coast, but Strabo may have had another homonymous place in mind, since he has just mentioned the Molochath River (modern Oued Moulonia), far to the west. The traditional site of Metagonion is almost due south of ("across from") Massalia, and it would be an important sailing marker for ships crossing the Mediterranean. If Strabo were referring to another place of the same name, it would be across from New Karchedon (modern Cartagena). The distances make little sense, although 3,000 stadia (about 375 miles) is close to the distance from traditional Metagonion to Massalia. Clearly the data have become confused at some time between Timosthenes and Strabo, but the material reflects faint vestiges of the former's account of the western Mediterranean.

34. Strabo, Geography *3.1.7 (W19, BNJ F16).*

To some, it [Kalpe] was a foundation of Herakles, among whom is Timosthenes, who says that it was called Herakleia in antiquity, and that its large circuit wall and shipsheds are still pointed out.

Commentary: Kalpe is identified with the northern Pillar of Herakles, modern Gibraltar, certainly one of the most prominent landmarks in the western Mediterranean. The account indicates that there was a city of the same name, whose location has not been found. This is the only citation of the earlier name Herakleia, but many places in the western Mediterranean were associated with Herakles and had been so since at least the fifth century BC and probably earlier (Herodotus, *Hist.* 2.33). Yet, as usual, Timosthenes reported on Kalpe from a seaman's perspective, remarking on its conspicuous walls and shipsheds. Kalpe is the most western location to appear in the fragments of *On Harbors.*

35. Stephanos, "Chalkeia" (BNJ F30).

Chalkeia, a Libyan city. [Alexandros] Polyhistor [F46] in Book Three of his *Libyka*: "Thus Timosthenes [?], whom Polybios [12.1.5] censures in his twelfth book, writing that in fact he is totally ignorant about Chalkeia, for it is not a city but copper mines."

Commentary: The manuscripts of Stephanos have "Demosthenes," rather than Timosthenes, which makes little sense and may be an error by the immediate source of the fragment, the polymath Alexandros Polyhistor of Miletos, active in the first century BC. He was quoting Polybios (a citation now lost except for this fragment), and if the reference is actually to Timosthenes, this is the only evidence that Polybios read him. Like Strabo, Polybios was generally highly critical of his geographical predecessors. The copper mines (*chalkeia*) in Libya may be those in Mauretania vaguely known to Strabo (*Geogr.* 17.3.11; see also Ptolemy, *Geographical Guide* 4.2.17).[20]

Religious and Cultic Writings

36. Scholia to Apollonios, Argonautica 3.847 (W41, BNJ F18).

Timosthenes in his *Exegetikon* agrees that Persephone is called Daira, as Aeschylus shows in his *Psychagogos* [F277 Radt].

Commentary: Of all the references to Timosthenes's work, this is most anomalous. The title *Exegetikon* is not documented elsewhere in the fragments, but seems to be a title for a cultic or religious work (Plutarch, *Nikias* 23.6). Daira, or Daeira ("Knowing One") was one of the Okeanids and the mother of Eleusis, a cultic name known in Athens by the fifth century BC and eventually associated with Persephone (Pherekydes [*FGrHist* #3] F45; Lykophron, *Alexandra* 710; Pausanias, *Descr.* 1.38.7). The fragment is unlikely to have come from *On Harbors*, but Timosthenes did write on cultic matters, and it and F37 may be the sole survivors of a work on that topic. Aeschylus's *Psychagogos* ("Spirit Raisers"), known from a handful of fragments, was about the death of Odysseus.

37. Strabo, Geography 9.3.10 (BNJ T3).

In addition to the cithara singers [at Delphi] there were flute and cithara players who did not sing but played a certain musical number called the Pythian Melody. It has five parts: the prelude, trial, encouragement, *iamboi* and dactyls,

---

20. Gabriella Ottone, "Strabone e la critica a Timostene di Rodi: un frammento di Polibio (XII.1.5) testimone del Περὶ λιμένων?," *Syngraphé* 4 (2002): 153–71.

and the pipes. Timosthenes, the naval commander of the second Ptolemy, composed this melody, and he also wrote *On Harbors* in ten books. In the melody he wished to commemorate the contest between Apollo and the serpent, having the prelude as a preface, the first attack of the contest as the trial, the contest itself as the encouragement, and the singing of the paean at the victory as the *iambos* and dactyl, with appropriate rhythms. One was suitable for hymns and the other, the *iambos,* for blaming, as in "to lampoon" [*iambizein*]. The pipes [*syringes*] imitate the death of the serpent, as if ending its life with its final hissing [*syrigmos*].

Commentary: F37 demonstrates the diversity of Timosthenes's writings, since it is clear that it does not come from *On Harbors*, and it was probably from the *Exegetikon* mentioned in F36. It is unlikely that he actually composed the Pythian Melody, which probably had been in existence since early times, but he may have revised it or published an authoritative version. The melody was a dramatization of the killing of the Python by Apollo, an event first documented in the *Homeric Hymn to Apollo* (355–374).

Hidden Fragment

38. Pliny, Natural History, *contents to Book 4.*

From the following authors . . . Timosthenes . . .

Commentary: In the index to Book 4 of the *Natural History*, Pliny listed Timosthenes as a source, but there is no mention of him in the book itself, and it remains speculative how he was used. Book 4 covers the northern Greek peninsula, various Mediterranean islands, the Black Sea region, and western Europe, all coastal regions (with the probable exception of extreme western Europe) that could have been discussed in *On Harbors*.

CHAPTER 3

The Politics of Cartography: Foundlings, Founders, Swashbucklers, and Epic Shields

Georgia L. Irby

IN ANTIQUITY (AS NOW) MAPS were powerful tools of propaganda.[1] They were exploited by the Milesian tyrant Aristagoras (Herodotus, *Hist.* 5.49), Aristophanes (*Clouds* 215–218), Julius Caesar, and Augustus, among others. Although pictorial maps are attested,[2] the surviving evidence is largely narrative, and cartography was expressed in a remarkable variety of literary genres. Geography quickly became the purview of both scientists and poets. We see robust geographical curiosity and awareness in the works of Homer and Pindar. Aeschylus's tragedies, especially the *Persians* and *Prometheus*, are replete with new geographical learning. Euripides's *Ion* is, at its heart, a tale of foundation and "colonization:" hence it is an expression of mapping at its most political. Callimachus included *periploi* in his Hymns to Artemis and Zeus (*Hymns* 1.4–27; 3.183–260), and Apollonios of Rhodes, in the *Argonautica*, draws on both Homeric traditions and contemporary advances in geography.

Maps reflect the prejudices and philosophies of those who make them. Geometrical aesthetics and philosophical beliefs dictate the shape of the entire world (circular or spherical) and its inhabited portion. Mapmakers often exaggerate the size and importance of their own countries, placing their homelands or a site of particular cultural importance at the center, providing ample details of familiar locales. The scale lessens and details attenuate for increasingly remote and unfamiliar places. While attacking Pytheas of Massilia (flourished ca. 340–290 BC), Strabo goes so far as to devalue the need to know about distant places (*Geogr.* 2.5.8), and Tacitus bluntly declares that, beyond a certain point, our information is "storied" (*fabulosa*; *Germania* 46.5).

The geocentric model of the cosmos was, in itself, a political choice. Aristarchos of Samos had proposed a heliocentric paradigm (Archimedes, *The Sand*

---

1. Daniela Dueck, *Strabo of Amasia: A Greek Man of Letters in Augustan Rome* (London: Routledge, 2000), 45.

2. E.g., Herodotus, *Hist.* 5.49; Aristophanes, *Clouds* 200–218; Diogenes Laërtios, *Lives* 5.51, for Theophrastus's bequest of a map.

Reckoner), explaining celestial phenomena with elegant simplicity, thus obviating the need for complicated Eudoxan planetary motions.[3] But the theory was fraught with controversy, and Kleanthes had even proposed charging Aristarchos with impiety for "disturbing the hearth of the universe because he sought to save the phenomena by assuming that the heaven is at rest while the earth revolves along the ecliptic and at the same time rotates about its own axis" (Diogenes Laërtios, *Lives* 7.174).[4] On a geocentric model, humans inhabit the center of the universe. The Pythagorean model situated the earth just off center on ethical grounds: flawed humanity is not sufficiently worthy to occupy the very center of the cosmos (Philolaus, *TEGP* 24). The Pythagorean hypothesis, although not universally accepted, hardly represented a violent reorganization of the celestial sphere. Heliocentrism, however, was a radical deviation that removed mankind from (near) the center of the universe and thus diminished human status. Heliocentrism, consequently, undermined the Greek sense of self worth, rendering humankind insignificant, like a "fleck of stellar dust somewhere in a boundless vastness, which will end in freezing darkness with the last visible star fading from sight as the universe flies apart."[5]

Homer

Many thinkers explored the relationship of the earth to the cosmic sphere, and in Homer we find an early attempt at mapmaking that shows this deployment of earth and heavens. The opulent artwork on Achilles's shield depicts a synthesized microcosm of Homer's world, both terrestrial and celestial (*Il.* 18.478–608). This Homeric shield-map emphasizes the human occupation of the cosmos. On Achilles's bronze-plated shield is the earth in its relationship to the major celestial bodies—Orion, the Hyades, Pleiades, and Ursa Major (*Il.* 18.483–489):[6]

3. Heliocentrism was largely rejected until revived by Copernicus (1473–1543). In his old age Plato had considered its merits, and Seleukos (flourished 165–135 BC) advanced the theory (Plutarch, *Platonic Questions* 8.1; Duane W. Roller, "Seleukos of Seleukia," *Antiquite Classique* 74 [2005], 112–14), as, seemingly, did Metrodoros and Krates (Pseudo-Plutarch, *Placita* 2.15), and perhaps Anaximander, though the evidence is contradictory.

4. Cf. Tracey Elizabeth Rihll, *Greek Science* (Oxford: Oxford University Press, 1999), 77–78; G. E. R. Lloyd, *Greek Science After Aristotle* (New York: Norton, 1973), 54.

5. Rihll, *Greek Science*, 73. The theory belied both common sense (Aristarchos had theorized that the stars had to be sufficiently distant to account for the lack of stellar parallax) and Aristotelian physics. Aristotle argued that heavy objects tend towards the center. Once an object reaches its natural place, with nowhere else to go, it stays at rest. Earth, the heaviest element, is not only at rest but cannot be moved except by a force great enough to overcome natural tendency (*On the Heavens* 296b26–297a7).

6. See Philip R. Hardie, "Imago Mundi: Cosmological and Ideological Aspects of the Shield of Achilles," *JHS* 105 (1985), 11.

On it, (Hephaestus) made the earth, on it the sky, on it the sea, both the untiring sun and the full moon, on it all the constellations which encircle the sky, the Pleiades and Hyades, and the strength of Orion, and Arktos the bear, she whom they also call the "Wagon," she who is turned about the axis, she who narrowly watches Orion, she who alone is without a share of Ocean's bathing pool.

Pherekydes suggests a similar cosmic arrangement in *Heptamychos*.[7] The earliest maps are, in fact, celestial, and the Lascaux cave paintings (15,000 years ago) may feature star charts and representations of constellations.[8] Such data contribute to the regulation of political and religious calendars, and they advance cartographic aims, including determining latitudes and estimating the circumference of the earth.

Homer's earth comprises two cities, one of peace, one of war, together illustrating the variety of human experience.[9] Around the "outermost rim" (*Il.* 18.608: ἄντυγα πὰρ πυμάτην) flows the "great strength of the river Ocean" (*Il.* 18.607: ποταμοῖο μέγα σθένος Ὠκεανοῖο). Omitting topography and natural landmarks, this "map" instead emphasizes how human civilization fits within the cosmos, showing the deployment not of individual places but of human political institutions (the cities of peace and war) to the celestial sphere. It is the river Ocean that limits this human settlement and achievement. Thus, the universe on Achilles's shield is not so much geocentric as it is androcentric, anticipating Aristotle's maxim that "man is by nature a political animal" (*Politics* 1253a2: ὁ ἄνθρωπος φύσει πολιτικὸν ζῷον)—that is, a creature who dwells in a self-sufficient *polis* community.

The Milesians

The themes of Homer's world map are advanced by the Milesian philosophers. Anaximander of Miletos (ca. 555 BC) is the first Greek to be credited in the formal sense with drawing a map, "a geographical pinax" (Strabo, *Geogr.* 1.1.11: γεωγραφικὸν πίνακα). No doubt Anaximander's interest in cartography was

7. DK⁶ 7B2; Robert Hahn, *Anaximander and the Architects: The Contributions of Egyptian and Greek Architectural Technologies to the Origins of Greek Philosophy* (Albany: State University of New York Press, 2001), 204.

8. Michael Rappenglueck, "A Palaeolithic Planetarium Underground: The Cave of Lascaux," *Migration and Diffusion* 5 (2004), 93–119.

9. John B. Harley and David A. Woodward, *Cartography in Prehistoric, Ancient, and Medieval Europe and the Mediterranean* (Chicago: University of Chicago Press, 1987), 131; O. A. W. Dilke, *Greek and Roman Maps* (Ithaca, NY: Cornell University Press, 1985), 20, 145.

theoretical, but he may also have founded a settlement (perhaps on the Black Sea),[10] and his efforts may have been in answer to practical or political challenges. Anaximander's terrestrial map consisted in an outline (*perimetron*) of the earth and the sea that encircled the earth, a circumscribing stream of Ocean (Diogenes Laërtios, *Lives* 2.1–2; Agathemeros, *Geographias hupotuposis* 1.1 [Müller]).[11] Thus, Ocean confines both the Homeric world and Anaximander's earth. Anaximander's accompanying commentary probably included a cosmological introduction followed by an explication of the arrangement of natural and manmade landmarks, cities, and their attendant climates.[12] In this regard, Anaximander also follows Homer who set Achilles's two cities within the context of the cosmic sphere. In addition to *Circuit of the Earth*, Anaximander is also credited with at least one astronomical text, *On Fixed Stars*. His *Sphaira* (*Globe*) (Souda A-1986) likely treated the sphere of fixed stars, and not, as some have surmised, the construction of a terrestrial globe.[13]

No less than geocentrism, aesthetics and symmetry guided cartographic initiatives. For Anaximander the earth was a shallow, broad cylinder, its depth a third of its width, "like a stone column," hanging freely in the air, equidistant from other celestial objects (Anaximander, *TEGP* 19–20).[14] Details of Anaximander's *oikoumene*, however, are not extant, and any reconstruction of the map is at best speculative.[15] There is no consensus regarding the relative sizes of the landmasses,[16] or even of the map's center (Delphi, Delos, or Miletos).[17]

---

10. Hahn, *Anaximander and the Architects*, 202–3.

11. Dilke, *Greek and Roman Maps*, 23.

12. Dirk L. Couprie, Robert Hahn, and Gérard Naddaf, *Anaximander in Context: New Studies in the Origins of Greek Philosophy* (Albany: State University of New York Press, 2003), 49.

13. Harley and Woodward, *Cartography*, 134.

14. The evidence is fraught and contradictory: D. R. Dicks, *Early Greek Astronomy to Aristotle* (Ithaca, NY: Cornell University Press, 1970), 45–46.

15. Many scholars have assumed a division of landmasses into thirds, each separated by bodies of water: see Marcel Conche, *Anaximandre: Fragments et Témoignages* (Paris: Presses universitaires de France, 1991), 47, figure 2; J. Oliver Thomson, *History of Ancient Geography* (Cambridge: Cambridge University Press, 1948), 98, figure 11. Some distinguish between maps of the earth and those of the *oikoumene*: Conche, *Anaximandre*, 46; W. A. Heidel, *The Frame of the Ancient Greek Maps* (New York: American Geographical Society, 1937), 11–12. This tripartite division does not appear in the extant fragments of Anaxagoras and is not attested until the fifth century BC.

16. For equal division of the landmasses: Jeffrey M. Hurwit, *The Art and Culture of Early Greece: 1100–480 B.C.* (Ithaca, NY: Cornell University Press, 1985), 208; Couprie, *Anaximander in Context*, figure 3.16. For an unequal allotment of the landmasses: Robert S. Brumbaugh, *The Philosophers of Greece* (Albany: State University of New York, 1964), 22; Conche, *Anaximandre*, 47, figure 2; Thomson, *History of Ancient Geography*, 98, figure 11.

17. For Delos at the center: Hurwit, *Art and Culture*, 208; Delphi: Brumbaugh, *The Philosophers of Greece*, 22; Harley and Woodward, *Cartography*, 135; Couprie, *Anaximander in Context*, 194; or Miletos or Didyma: Conche *Anaximandre*, 46; Christian Froidefond, *Le mirage égyptien dans la littérature grecque d'Homère à Aristote* (Paris: Ophrys, 1971), 167.

The shape of Anaximander's inhabited region is also disputed.[18] The map may have incorporated a three-point coordinate system, corresponding to the rising and setting of the sun on the days of the equinoxes and solstices.[19] Most ancient maps were more or less symmetrical,[20] and Herodotus of Halicarnassus (ca. 450 BC) states that such circular maps appear to have been drawn with a compass or turned on a lathe (*Hist.* 4.36.2: ἀπὸ τόρνου).[21] Anaximander's map likely responded to Near Eastern cartographic philosophy, where maps represented "order imposed on the physical landscape, or the idea of the physical landscape, both terrestrial and celestial."[22] This worldview also accords

18. For the *oikoumene* as a circle: Brumbaugh, *The Philosophers of Greece*, 22; Conche, *Anaximandre*, 47; Hurwit, *Art and Culture*, 208; a parallelogram: John L. Myres, *Herodotus: Father of History* (Oxford: Clarendon, 1953), 6, figure 5; Heidel, *The Frame of the Ancient Greek Maps*, 1, figure 1; or as a parallelogram inscribed in a circle: Thomson, *History of Ancient Geography*, 97, figure 10.

19. Hahn, *Anaximander and the Architects*, 204, following Heidel, *The Frame of the Ancient Greek Maps*, 17–20, 57. The equinoctial and solsticial points are determinable with a seasonal sundial. Diogenes Laërtios, *Lives* 2.1–2, credits Anaximander with "discovering" the equinoxes and solstices as well as the gnomon, and for applying the gnomon to sundials at Sparta. The gnomon enables the time of day to be estimated trigonometrically. Otto Neugebauer and R. A. Parker, *Egyptian Astronomical Texts I: The Early Decans* (London: Lund Humphries, 1960), 116–21, have shown that the origins of the seasonal hours derive from the Egyptian practice of marking nighttime according to decanal risings. The Babylonians and Egyptians were familiar with sundials, and the Greeks may have learned about sundials from the Babylonians, as Herodotus claimed (*Hist.* 2.109.3). Anaximander can hardly have invented the gnomon. The Greeks were estimating time of day by shadow tables, which utilized gnomons, well before Anaximander, but a lack of archaeological data coupled with ambiguous literary evidence prevent resolution: see Sharon L. Gibbs, *Greek and Roman Sundials* (New Haven: Yale University Press, 1976), 6. The Milesian philosopher is, however, credited with incredible intellectual feats, among which are the discovery of the obliquity of the zodiac (Pliny, *Nat. Hist.* 2.31). Nonetheless, Anaximander may simply have been the first Greek to make scientific use of the gnomon, perhaps learning its applications from the Egyptians as Thales is said to have learned geometry from them: see William A. Heidel, "Anaximander's Book: The Earliest Known Geographical Treatise," *Proceedings of the American Academy of the Arts and Sciences* 56 (1921), 244.

20. Anaximander (Agathemeros, *Geographias hupotuposis* 1.1 [Müller 2.471]; Diogenes Laërtios, *Lives* 2.1–2; Herodotus, *Hist.* 2.109); Hekataios (Agathemeros, *Geographias hupotuposis* 1.1 [Müller, 2.471]); Eratosthenes (Strabo, *Geogr.* 1.4.1–9); Krates (Strabo, *Geogr.* 2.5.10).

21. Cf. Plato's model of the universe in the myth of Er: *Republic* 10.14. Plato employs the analogy a spindle whorl in order to explain the geocentric cosmos as the orbits of the celestial bodies. The description may be cartographically motivated: Harley and Woodward, *Cartography*, 138.

22. Francesca Rochberg, "The Expression of Terrestrial and Celestial Order in Ancient Mesopotamia," in *Ancient Perspectives: Maps and their Place in Mesopotamia, Egypt, Greece and Rome*, ed. Richard J.A. Talbert (Chicago: University of Chicago Press, 2012), 11. An extant clay tablet shows a circular map of the "New Empire," a world framed by the "Bitter River" and dotted with significant topographical details: a mountain, canal, swamp, and cities: Rochberg, "Terrestrial and Celestial Order in Ancient Mesopotamia." See Kurt A. Raaflaub and Richard J. A. Talbert, *Geography and Ethnography: Perceptions of the World in Pre-Modern Societies* (New York: Wiley, 2009), 147.

with the precepts of pre-Socratic cosmogonies which sought to explain the world as an ordered system (cosmos) emerging from chaos on the Hesiodic model.

Cartographic details were slowly fleshed out by Greek inquiry. Anaximander's fellow Milesian, Hekataios (ca. 500 BC) was the first Greek to attempt a systematic description of the world in his geographical treatise *Periodos* (or *Periegesis*) *Ges* (journeying around the earth),[23] likely a copy (Strabo, *Geogr.* 1.1.1) or improvement of Anaximander's terrestrial map.[24] By the fifth century BC, firmly established is the paradigm of a tripartite world with three landmasses, each separated by waterways: the Nile borders Libya and Asia, the Phasis River and Euxine (Black or Pontic) Sea divide Asia from Europe, and the Mediterranean Sea delimits Europe from Libya (Herodotus, *Hist.* 4.45.2; cf., 4.36.2). Herodotus also describes a detailed map of the Greek world which the Milesian tyrant Aristagoras employed as a prop in 499/498 BC in his efforts to secure a Spartan alliance against Darius I, king of Persia.[25] The portable map, probably in bronze, was engraved on a *pinax*, as was Anaximander's map. Aristagoras showed this map, a circular "going-round" or *periodos*, to the Spartan king Kleomenes:[26] "a bronze tablet, on which the circuit of all the earth was engraved, and all the sea and all the rivers" (Herodotus, *Hist.* 5.49.1). Aristagoras left Sparta without securing an alliance. Kleomenes had concluded that the distance ("about three months") rendered the threat too remote for immediate action. Herodotus, nonetheless, demonstrates the political potential of maps, the physical artefacts. Aristagoras believed that the Spartans might take the Persian threat more seriously on the strength of a graphic, and he emphasized the Persian domination over the lands that he indicated to Kleomenes. The point is all the more emphatically (and comically) made by Aristophanes whose Strepsiades observed a map of the Greek world on display at Socrates's *Phrontisterion* during the Peloponnesian War. When Strepsiades noticed the proximity of Athens and Sparta, he vehemently demanded that the enemy *polis* be moved further away (*Clouds* 215–218).

23. Philip Kaplan, "Hekataios of Abdera" in *EANS*: 361. In criticizing circular maps, where he employs the term *periodos*, Herodotus may have had in mind Hekataios's map or one of its derivatives: Harley and Woodward, *Cartography*, 135.

24. Agathemeros proclaimed the new edition "more accurate so that it became a source of wonder" (*Geographias hupotuposis* 1.1, 2.461 [Müller]).

25. Dilke, *Greek and Roman Maps*, 23–24; Couprie, Hahn, and Naddaf, *Anaximander in Context*, 33.

26. Anaximander is alleged to have erected a sundial in Sparta (Anaximander, *TEGP* 1) and to have predicted an earthquake there (Cicero, *On Divination* 1.112).

Euripides's *Ion*

The propaganda value of maps—real and metaphorical—is further explored in the *Ion*, where Euripides examines a myth that provides a genealogical explanation of Athenian ethnic and political identity.[27] The geopolitical spotlight emphasizes human (Attic Greek) control of the *oikoumene*, its extent and its borders (like the river Ocean on Achilles's shield and Anaximander's *pinax*).[28] The *Ion* is a myth of foundation and settlement, whereby the contours of the *oikoumene* are brought into focus by the exploration of an unfolding geographical curiosity and awareness. We learn from Hermes at the play's beginning that Ion received his name from Apollo. The boy, "going forth" (ἰών), will be renowned through Greece as the founder of ancient cities (74–75).[29] Hence, geography— that is to say, settlement—is Ion's purpose. By "going forth," Apollo's son establishes settlements, spreads Athenian hegemony, and expands the outlines of the Athenian and Apolline world, whose center is emphatically positioned at Delphi (5–6, 223, 462: ὀμφαλὸν μέσον). The play is rich in visual detail, a characteristic of epic, not tragedy "as we know it."[30] The play's emphasis on geographical features, especially in its attention on travel and settlement, is another trajectory more characteristic of epic than tragedy.[31]

27. Anne Pippin Burnett, "Human Resistance and Divine Persuasion in Euripides' *Ion*," *CP* 57 (1962), 89, 99–100, 101; Christian Wolff, "The Design and Myth in Euripides' *Ion*," *HSCP* 69 (1965), 169–94; George B. Walsh, "The Rhetoric of Birthright and Race in Euripides' *Ion*," *Hermes* 106 (1978), 301–15; Barbara Elizabeth Goff, "Euripides' *Ion* 1132–1165," *PCPhS* 34 (1988), 42; Katerina Zacharia, *Converging Truths: Euripides' Ion and the Athenian Quest for Self-definition* (Leiden: Brill, 2003), 60–65.

28. It is beyond our scope to explore the ramifications of Athenian autochthony, for which see Henri Grégoire, ed., *Euripide: Tome III* (Paris: Les Belles Lettres, 1950); Wolff, "The Design and Myth," 174–76; Donald John Mastronarde, "Iconography and Imagery in Euripides' *Ion*," *CSCA* 8 (1975), 163–76; Goff, "Euripides' *Ion* 1132–1165."

29. According to Carl A. P. Ruck, "On the Sacred Names of Iamos and Ion: Ethnobotanical Referents in the Hero's Parentage," *CJ* 71 (1976), 235–52, the etymology of Ion's name is analogous with Iamos, derived from the imagery of his birth scene (cf., Pindar, *Olympian* 6): ἰός, poison or venom, and ἴον, flowers. Contradicting the playwright's own etymology (831: Ἴων, ἰόντι δῆθεν ὅτι συνήντετο), Ruck, "On the Sacred Names," 236, argues that "the entire naming episode in the play is a burlesque of misunderstandings and the poet seems to have structured the narrative of the tragedy to hint at a meaning which would be more appropriate to Ion's sacred nature." Granted, poison is a central image in the play. Nonetheless, as Mastronarde, "Iconography and Imagery," 17, writes "no single approach can exhaust the richness of a good Euripidean play," and the protagonist's name may very well work on multiple registers. Euripides was quite willing to magnify a simple etymology, and there is no reason to reject that interpretation in the context of the play's rich geographical grounding.

30. Wolff, "The Design and Myth," 180.

31. The play's eponymous protagonist demonstrates a high level of geographical knowledge and curiosity. As Kreusa describes Athens, Ion inquires about the "Long Rocks" (Μακραί: 283), a question at its most basic level, that serves as a literary plot device. As the site of Apollo's rape of Kreusa and where the teenage girl abandoned her baby, the Long Rocks are a central topographical

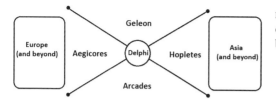

FIGURE 3.1. The *oikoumene* of Euripides's *Ion*, drawing by author.

At the play's end Athena stresses Ion's geopolitical role. Ion's lineage will dominate Greece (1575–1594). "His four sons, from a single root (i.e., Ion), shall be the eponymous founders of the Ionic tribes: Geleon, Hopletes, Argades, and Aegikores. In time their own children will go forth, settling the Cycladean island-cities which strengthen the Attic lands. And then they shall settle the lands towards Asia and those towards Europe" (1586–1587: Ἀσιάδος τε γῆς Εὐρωπίας τε). And these people, the Ionians after Ion, will have renown (κλέος) by the grace of Ion's name (1587–1588: τοῦδε δ᾽ ὀνόματος χάριν Ἴωνες ὀνομασθέντες ἕξουσιν). Ion's descendants, the Ionians, will inhabit the eastern Aegean, Asia, and Europe.

The story is a foundation myth for all of Greece. Athena tells us that Kreusa's heirs will establish Greek Attic-Ionian geopolitical mastery over the entire Greek-speaking world. Ion's legacy, which includes Europe, Asia, and the Aegean, completes the map of the Greek world. The process of settlement and its articulation in the play reflect Greek scientific accounts of ordering the world for the purpose of understanding and controlling it. Hence, Euripides's *Ion* becomes an expression of Apolline order and the god's authority over the inhabited world, just as Vergil's *Aeneid* is, among other things, an expression of Augustan control over the Roman Empire.

Ion's map, however, requires a frame, a limit, like Ocean in Homer and Anaximander. Euripides perfects the contours of Ion's world map by specifying its edges. Determined to learn his mother's identity, and employing the verb which yields his name (ἐπ-ελθὼν) Ion swears that he will be "going out" to search for her "towards Asia and Europe's boundaries" (1356: Ἀσιάδ᾽ Εὐρώπης θ᾽ ὅρους). He will travel to the eastern and western limits of the *oikoumene* and beyond.[32] The tension between order and violence is sustained in these

---

image of the play. The Long Rocks are also a prominent feature of the Athenian topography of the *Ion* about which the title character already knows something. A little later, Kreusa explains how she came to marry a foreigner, Xouthos, as a prize for his military aid in vanquishing Euboea, some neighboring land (τις γείτων πόλις). Ion immediately interjects "with watery boundaries" (ὅροις ὑγροῖσιν: 295), showing that he knows that Euboea is an island.

32. See Glen W. Bowersock, "The East-West Orientation of Mediterranean Studies and the Meaning of North and South in Antiquity," in *Rethinking the Mediterranean*, ed. William V. Harris

liminal areas.[33] Also emphasized is the process of civilization that emerges from savagery through the imagery of liminal monsters, as in Hesiodic and pre-Socratic cosmogonies where civilization evolves from natural forces: Herakles vanquishes the Lernean Hydra (190–192);[34] Bellerophon overcomes the Chimera (201–204). The Gorgon (989, 1421–1423) and the Gigantomachy (206–218) are also described,[35] and Kreusa is assimilated to Echidna (1261–1265). As such, the *Ion* is a statement of Athenian self-assertion and control of empire, supported by the narrative map of Athenian hegemony provided within the text.

Like Homer, Euripides includes a celestial map. The major celestial bodies are woven into the tapestries plundered from the Amazons by Herakles and displayed by Kreusa's husband Xouthos at his banquet (1146–1158).[36] The heavens are presented as a kinetic textual map. In heaven's circle, Ouranos musters the stars; Helios traverses the sky from east to west, the cardinal directions emphasized elsewhere in the play. The stars keep pace with black-robed night in her chariot, evoking the appearance that the "fixed stars," including the Pleiades and Orion, move as a unit through the night sky. Ursa Major turns her "golden tail within the vault" as she rotates around the Pole stars. The full moon divides the month; then "light's herald," Dawn, vanquishes the stars. Euripides moves chronologically through the sky, from evening to night to morning. And he recounts the systematic arrangement and behavior of the sun, moon, and constellations, that is, the regular succession of night and day. Thus the poet describes an ordered cosmology.[37] This mathematical precision of the celestial sphere extends to the terrestrial realm as civilization emerges from the disorder and raw nature of liminal regions, and as Ion (sent by Apollo) promulgates knowledge and understanding and control of the growing Apolline universe.

Apollonios's *Argonautica*

As in the *Ion*, we see the theme of geopolitical hegemony sustained in Apollonios of Rhodes's *Argonautica*, a Hellenistic recounting of the Homeric-like adventures of Jason in his quest for the Golden Fleece. Unlike Homer, whose geography includes fantastical and legendary places, Apollonios interprets

---

(Oxford: Oxford University Press, 2005), 167–78, for the primacy of an east/west division of the Mediterranean.

33. Wolff, "The Design and Myth," 176–77; Mastronarde, "Iconography and Imagery," 165–66; Goff, "Euripides' *Ion* 1132–1165."

34. Mastronarde, "Iconography and Imagery," 166.

35. Mastronarde, "Iconography and Imagery," 168.

36. See Goff, "Euripides' *Ion* 1132–1165," 43–44.

37. Cf. Heraclitus, *TEGP* 89 and Parmenides, *TEGP* 10 where justice and morality determine the regularity of the stars: Goff, "Euripides' *Ion* 1132–1165," 52n2.

Jason's journey, which includes real and imagined landscapes, within the parameters of the burgeoning knowledge of contemporary scientific geography.[38] Likewise evident are the themes of settlement and hegemony over the *oikoumene* that we have noted in Euripides's *Ion*.

Apollonios's geopolitical trajectory is neatly evident in the ecphrasis of Jason's cloak, a gift from his divine patroness Athena (*Argon.* 1.730–767). On the cloak are seven scenes: (1) the Cyclopes forge thunderbolts (1.730–734);[39] (2) Amphion and Zethus build Thebes (1.734–741); (3) Aphrodite admires herself in Ares's shield (1.742–746); (4) in the center, the sons of Elektryon (Herakles's grandfather) fight the marauding Taphians who will kill Alkmene's brothers (1.747–751);[40] (5) a chariot race features Pelops and Oenomaus, who cheats despite his divinely fast horses (1.752–758); (6) Apollo kills Tityos, who had raped Leto (1.759–762); and (7) Phrixos chats with the golden ram, who had conveyed him to safety from the Hellespont (1.763–767). According to the scholiast, the scenes—unified in purpose—are illustrations of divine intent as demonstrated through civilization on earth (1.763–764a).[41] The ecphrasis invites other readings: the scenes may refer to the "proper roles" of men and women,[42] underscore the epic's subtext of love and hate,[43] offer a "symbolic representation of the journey,"[44] or they may be completely lacking in thematic unity.[45] The scenes together are specific, distant, and mythological, emphasizing not martial themes (as common in epic) but an eventual success that is peaceful and civilized.[46] Shapiro contends that these scenes were included in accord with the aesthetic principles of contemporary art rather than for any unifying symbolism.[47]

38. Dag Øistein Endsjø, "Placing the Unplaceable: The Making of Apollonios' Argonautic Geography," *GRBS* 38 (1997), 373–85; cf. André Hurst, "Géographes et poètes: Le cas d'Apollonios de Rhodes," in *Sciences exactes et sciences appliquées à Alexandrie*, ed. Gilbert Argoud and Jean-Yves Guillaumin (Saint-Étienne: Publications de l'Université de Saint-Étienne, 1998), 279–88.

39. A scene deliberately imitated by Vergil, *Aen.* 8.424–427. See Peter Green, trans., *Apollonios of Rhodes: The Argonautika* (Berkeley: University of California Press, 2007), 217.

40. See Francis Vian and Émile Delage, *Apollonios de Rhodes: Argonautiques* (Paris: Belles Lettres, 1974), 258.

41. Green, *The Argonautika*, 216.

42. John F. Collins, "Studies in Book One of the *Argonautika* of Apollonios Rhodius" (PhD diss., Columbia University, 1967), 83.

43. Charles Rowan Beye, *Epic and Romance in the* Argonautika *of Apollonios* (Carbondale: Southern Illinois University, 1982), 91–92.

44. James J. Clauss, *The Best of the Argonauts: The Redefinition of the Epic Hero in Book One of Apollonios'* Argonautica (Berkeley: University of California Press, 1993), 128; Green, *The Argonautika*, 217.

45. Donald N. Levin, "*Diplax Porphuree*," *RFIC* 93 (1970), 17–36.

46. See Levin, "*Diplax Porphuree*"; Carol Una Merriam, "An Examination of Jason's Cloak (Apollonios Rhodius, *Argonautika* 1, 730–768)," *Scholia* 2 (1993), 70.

47. H. Alan Shapiro, "Jason's Cloak," *TAPhA* 110 (1980), 263–86. Cf. Green, *The Argonautika*, 216. Robert D. Williams, trans., *The Aeneid of Vergil* (Basingstoke, UK: MacMillan, 1981), ad loc., makes a similar observation regarding Aeneas's shield; cf. David Alexander West, "*Cernere erat:*

Lawall considers the cloak didactic, arguing that its scenes offer disconnected object lessons: in hubris and piety, the preeminence of skill over brute force, the conquest of love over war, and the justification of treachery.[48] Merriam denies that the cloak has any didactic purpose, but she regards the discrete scenes as illustrating elements of the epic as a whole by giving "alternatives to heroic violent action which will bring success to Jason and his crew."[49] In other words, the cloak foreshadows the methods that the Argonauts will employ in order to accomplish their goals.

We do not argue that the scenes on Jason's cloak in any way constitute a map, real or metaphorical. But some of the individual scenes do emphasize the themes of origin, foundation, and settlement that were foregrounded in the *Ion*. The cloak explicitly shows the founding of Thebes. It was a princess of Thebes, Dionysus's aunt Ino, who set into motion the events that culminated in Jason's quest. Because of her jealousy of her stepchildren Phrixos and Helle, Ino promulgated a false oracle from Delphi (that human sacrifice would dispel a lingering drought). This in turn induced Nephele, the cloud-like mother of Phrixus and Helle, to send the magical flying ram to save her children from their evil stepmother (Apollodoros, *Library* 9.1). Jason's adventure, in fact, begins in Thebes well before his birth. In its cattle raid and chariot race scenes, Jason's cloak alludes to other founders. Ruling in the Argolid was Herakles's maternal grandfather Elektryon, a son of Andromeda and Perseus (the legendary founder of Mycenae; Pausanias, *Descr.* 2.25.9; 15.4; 16.3–6; 18.1).[50] According to folk etymology, another son of Perseus, Perses, gave his name to the Perseid dynasty and to the "Persians" (Herodotus, *Hist.* 7.61.3; 150.2). Pelops, likewise, is the eponymous hero of the Peloponnese ("the island of Pelops") whose cult at Olympia developed into a foundation myth for the Olympic games, the most important of the pan-Hellenic games and perhaps the most profound expression of "Greekness." Finally Apollo is the Greek god of "colonization."[51] Among other things, Apollo's sanctuary at Delphi, which exerted "nothing short of a missionary influence,"[52] mandated the establishment of numerous Greek settle-

---

The Shield of Aeneas," *PVS* 15 (1975–1976), 1–6, who notes that the rich detail and references to color, texture, and the position of the scenes show that Vergil intended the scenes as conceivable and effective on an actual artifact.

48. Gilbert Lawall, "Apollonios' *Argonautika*: Jason as Anti-Hero," *YCS* 19 (1966), 121–69. Cf. Hyginus, *Fabulae* 84 for Pelops's bribery of Myrtilus; compare Medea's betrayal of her own brother Apsyrtus: Apollonios, *Argonautica* 4.303–16; Green, *The Argonautika*, 218.

49. Merriam, "An Examination of Jason's Cloak," 72.

50. Alkmene had refused to consummate her marriage with Amphytrion until he avenged the deaths of her brothers, killed by the Taphians.

51. William G. G. Forrest, "Colonization and the Rise of Delphi," *Historia* 6 (1957), 160–75.

52. Walter Burkert, *Greek Religion*, trans. John Raffan (Cambridge: Harvard University Press, 1985), 143.

ments. Tellingly, many cities were named for Apollo (see Thucydides, *Hist.* 6.3.1).[53]

The allusions are subtle and erudite. Nor do all the elements of the cloak ecphrasis contribute to Apollonios's geopolitical trajectory. But, like Hellenistic literature in general, replete with obscure references, the *Argonautica* is a dense work that operates on multiple levels.[54] Just as the *Ion* sketches the Attic world in the broadest strokes, so too does the cloak ecphrasis in the *Argonautica* present an impressionistic outline of Apollonios's Jasonian world. Apollonios presents the Peloponnese as the western extent of Greek culture, its center at Thebes, and its eastern edge at the Hellespont on the Black Sea, to where the ram carries Phrixos and Helle, the gateway of the "Persian" Empire (Herodotus, *Hist.* 7.6, 33–37) and to Jason's expedition (Apollonios, *Argon.* 2.548). The scenes on the cloak, taken together, emphasize lawful Greek society in contrast with the lawlessness of the "other:" the obedient Cyclopes; the piratical Taphians; Oenomaus's fear of being overthrown by a son-in-law (Apollodoros, *Library* 2.4),[55] Tityos's attempted rape of Leto, and Aeëtes's xenophobia, insinuated by account of the golden ram. Apollonios manipulates the Jason myth into a coherent narrative whose aim is to complete a circumnavigation of the world. Jason's expedition and his encounters with peoples on its fringes establish order and impose Greekness over the growing *oikoumene*. Thus, Jason is here a doublet for Alexander, whose aim was to expand the influence of Greek culture and to unite the world.

Geography in Vergil

The geopolitical trajectories evident in Homer, Euripides, and Apollonios of Rhodes are palpable also in Vergil, whose debt to his Greek models, especially Homer and Apollonios of Rhodes, cannot be overestimated. Aeneas's shield is a rich intertextual ecphrasis, drawing from the Pseudo-Hesiodic "Shield of Herakles,"[56] Jason's cloak, and Achilles's Iliadic shield.[57] Vergil's shield ecph-

53. Burkert, *Greek Religion*, 144.

54. Green, *The Argonautika*, 219, observes: "I sometimes suspect that Apollonios picked up his illustrations very much at random, with a view to puzzling the pedants, much as James Joyce worked in obscure Dublin references throughout *Ulysses* in order (he said) to keep the scholars busy."

55. Oenomaus's fear parallels Danaus's fear of his grandson Perseus: each tries to prevent his daughter from lawful marriage.

56. Riemer Anne Faber, "Vergil's Shield of Aeneas (*Aeneid* 8.617–731) and the *Shield of Herakles*," *Mnemosyne* 53, no. 1 (2000), 48–57.

57. Robert Cohon, "Vergil and Pheidias: The Shield of Aeneas and of Athena Parthenos," *Vergilius* 37 (1991), 22–30, suggests that Vergil knew of and drew from Phidias's shield of Athena Parthenos.

rasis both replicates and self-consciously diverges from the themes and *topoi* of the Homeric passage[58] and its other models.

By the end of the first century BC in Rome, political and geographical achievements were synthesized, and the size and shape of the physical world "was becoming almost synonymous with Roman imperial aspirations."[59] Geography showed the extent of Roman authority, and the poets emphasized the breadth of Roman hegemony by playing up the borders of the known world.

Throughout the *Aeneid*, Vergil deftly manipulates geography. The epic is, at its core, a tale of travel (like the *Odyssey* and *Argonautica*) and settlement (like the *Ion*).[60] Aeneas treks from the eastern edge of the Greco-Roman world (Troy, whose harbor was a strategic Bronze Age gateway of Asia Minor) to found a new home at the (then) western extremity, Hesperia. Along the way Aeneas receives enigmatic directions and advice from a host of well intentioned Trojan specters and divine agents, including the ghosts of the immodest Hector, Priam's best son (Vergil, *Aen.* 2.289–295), and Creusa, Aeneas's deceased wife. Creusa offers a vague toponym, Hesperia, that refers to the western lands of the evening star (Spain or Italy) and is reminiscent of Hesiod's Hesperides (*Theog.* 215, 275, 518). In Hesperia, the river Thybris flows in a gentle stream, which Aeneas will reach after a long sea journey (Vergil, *Aen.* 2.780–782: *terram Hesperiam venies, ubi Lydius arva inter opima virum leni fluit agmine Thybris*). Apollo's oracle at Delos enjoins Aeneas to "seek out your ancient mother" (3.94–98: *antiquam exquirite matrem*), a proclamation falsely interpreted by Anchises as referring to Crete. Finally the Penates, reiterating Creusa's long-forgotten mandate, specify the new homeland as a place which "the Greeks call by name Hesperia" (3.163–168: *Hesperiam Grai cognomine dicunt*).

Eventually, explicit cartographic advice comes from fellow Trojan exile, the prophet Helenus (3.410–432). Helenus tells Aeneas "when the wind moves you to the Sicilian shore and the Pelorian straits open up the narrows, may you aim for the land to port and the open waters to your left in your course. Avoid the starboard shore and waves." Helenus explains how Sicily was once attached to Italy. But "with force and waves, the sea cleft the Hesperian flank from Sicily and, between them, with a narrow tide it washed fields and cities drawn apart from the shore" (cf. Strabo, *Geogr.* 1.3.10). As this passage and

---

58. Michael C. J. Putnam, *Vergil's Epic Designs: Ecphrasis in the Aeneid* (New Haven: Yale University Press, 1998), 136–37, 167–69; cf. Lois V. Hinckley and Michelle Thorne, "The Shields of Achilles and Aeneas in Dialogue," *NECN* 21 (1993–1994), 149–55.

59. Katherine Clarke, *Between Geography and History: Hellenistic Constructions of the Roman World* (Oxford: Clarendon, 1999), 191–92.

60. The very point of the *Aeneid* is the founding of Rome (e.g., *Aen.* 1.257–296; 4.223–237; 8.729–731). Along his journey, Aeneas meets other emigrants, including Dido, a Tyrian exile, Evander and Diomedes, who were among those Greeks who had established settlements in Italy, as well as his fellow Trojan exiles Helenus, founder of Buthrotum, and Acestes, king of Sicily.

others demonstrate, Vergil was profoundly knowledgeable about the contemporary sciences, including geography.[61] In this graphic, verbal map, Helenus relays crucial cartographic data. He expounds on maritime dangers, including sea monsters (Scylla) and whirlpools (Charybdis): "And from the deepest abyss of the whirlpool, thrice (daily) Charybdis sucks in the waves and again raises them up alternately under the breezes, and she beats the stars with a wave. But a cave tucked away in a blind lair restrains Scylla from thrusting forth her faces and drawing ships to the rocks."[62] Helenus's verbal map is comparable with the narrative geographies included in Strabo, Mela, the elder Pliny, and others. No account of Sicily, for example, omits the dangerous whirlpool.[63]

The geographical accounts of Sicily also repeat the Scylla story. The strait between Italy and Sicily is particularly dangerous, and where the poets provide mythical explanations (a sailor-snatching sea-monster), geographers offer more rational exegeses. Strabo cites Polybios who compares the mythic Scylla with predatory fishes (*Geogr.* 1.2.14). Like Polybios's *galeotes* (swordfish or dogfish), Scylla "fishes there for dolphins or dogfish, or whatever larger fish she could take anywhere" (Homer, *Od.* 12.95–97). Scylla hunts dolphins, dogfish, and sailors there; there, the *galeotes* hunt tuna; and men in turn hunt the *galeotes* and their valuable swords.

With clear, precise directions, Helenus provides useful information about the course of travel as well as the dangers to expect when approaching western Italy. In fact, Helenus has given Aeneas a map. In content and tone, Helenus's map resembles the narrative maps rendered in more "scientific" sources, and it includes the same important data: the relationship between two places, Italy and Sicily; maritime dangers, sea monsters and whirlpools; geological history, how Sicily broke off from Italy; and even "history" or mythology. In its verbal

---

61. The invocation of the Muses in Vergil's *Aeneid* includes an explication of the spatial deployment of Rome and Carthage, where Carthage and the mouth of the Tiber are opposite one another (1.13–14: *Carthago, Italiam contra Tiberinaque longe/ostia*). In this, Vergil reflects the view of Eratosthenes who placed Rome on the same meridian as Carthage: F65. See Martin Korenjak, "Italiam contra Tiberinaque longe/Ostia: Virgil's Carthago and Eratosthenian Geography," *CQ* 54 (2004), 646–49; Duane W. Roller, *Eratosthenes' Geography: Fragments Collected and Translated with Additional Material* (Princeton: Princeton University Press, 2010), 174, an error that Strabo is quick to correct (2.1.40) "although it [Karthage] is more to the west, but he [Eratosthenes] does not admit his excessive ignorance of these regions and of those on toward the west as far as the Pillars" (Roller).

62. Charybdis's power extends to the stars, but with violence, in stark contrast with the just empire of lawful rule that Aeneas's descendants will administer: *Aen.* 1.259–260.

63. Where the poets, however, say that Charybdis ebbs and flows three times a day, Polybios (*Histories* 34.3.10), Eratosthenes (Roller, *Eratosthenes*, 53), and Strabo (*Geogr.* 1.3.11) note two cycles. Strabo defends Homer, whom Vergil followed literally, by ascribing the erroneous triple gushing to either a copyist error (*Aen.* 1.2.16) or rhetorical hyperbole to induce greater fear (*Aen.* 1.2.36).

presentation, this map reflects a long tradition of cartographic data collected from a variety of written and oral sources, including accounts from sailors and other travelers.

Aeneas's Shield Map

Maps serve several purposes, both practical, like Helenus's map, and symbolic, like Achilles's shield. As powerful emblems of authority and order, maps systematize and establish human control over the world. In 46 BC Julius Caesar, who greatly valued the power of maps, commissioned a world map that was never completed.[64] Likewise recognizing the symbolic authority of maps, Augustus appointed Marcus Vipsanius Agrippa to continue Caesar's work, and he displayed that map prominently in the Portico of Vipsania.[65]

Like Agrippa's map,[66] Aeneas's shield represents the Roman world to show Augustus's control of it. A gift from his mother Venus, Aeneas's shield is certainly a summary of Roman history,[67] but it is also a map of the Roman Empire. The shield not only features important historical events and sites, but it also specifies the borders of Roman hegemony.[68]

Vergil's description is stunningly detailed. The shield is framed with dolphins around the edges, and the sea is shown between each scene (*Aen.* 8.671–674):

haec inter tumidi late maris ibat imago
aurea, sed fluctu spumabant caerula cano,
et circum argento clari delphines in orbem
aequora verrebant caudis aestumque secabant.

Between these scenes there went a golden image of the broadly-swollen sea, but the dark-blue depth foamed from its white-capped wave, and

64. *GLM* 21–55; *GRL* §1060. See further, Irby, "Tracing the Orbis Terrarum from Tingentera," this volume, chapter 4.

65. *GRL* §332–333.

66. James J. Tierney, "The Map of Agrippa," *PCA* 59 (1962), 26–27; Dilke, *Greek and Roman Maps*, 41–53. Cf., Irby, "Tracing the Orbis Terrarum from Tingentera."

67. Stephen J. Harrison, "The Survival and Supremacy of Rome: The Unity of the Shield of Aeneas," *JRS* 87 (1997), 70–76; Andreola Francesca Rossi, "*Ab urbe condita*: Roman History on the Shield of Aeneas," in *Citizens of Discord: Rome and its Civil Wars*, ed. Brian W. Breed and Cynthia Damon (Oxford: Oxford University Press, 2010), 145–56; D. E. Eichholz, "The Shield of Aeneas: Some Elementary Notions," *PVS* 6 (1966–1967), 45–49.

68. Ida Östenberg, "Demonstrating the Conquest of the World: The Procession of Peoples and Rivers on the Shield of Aeneas and the Triple Triumph of Octavian in 29 B.C.," *Orom* 24 (1999), 155–62.

around dolphins, bright with silver, struck the level waters with their tails
and plowed the spray.

Thus Vergil echoes Homer as well as early Hellenic and Near Eastern conceptual maps where Ocean (here Vergil's *delphines*) encircles the disc of the earth.
Despite advances in geographical knowledge, the circular, symmetrical
map-paradigm had endured. We assume that the shield was circular, or at least
oval: Vergil specifies only that the fabric of the shield is indescribable (*Aen.*
8.625: *clipei non enarrabile textum*). Herodotus, a little more than a century
after Anaximander, derided the circular paradigm in no uncertain terms: "And I
laugh (γελῶ) seeing that many have before now scratched out circuits of the
world, and not one is rendered reasonably; for they describe Ocean, made round
as if on a lathe, flowing all around the earth, and showing Asia equal to Europe"
(*Hist.* 4.36.2). A century after Herodotus, Aristotle agreed. A circular representation of the earth, in Aristotle's opinion, was theoretically impossible given
the sphere's geometry and empirically impractical due to the ratio of the *oikoumene's* width to its breath (*Meteorology* 2.5.362b.13). Geminos (speculatively,
though not improbably, first century BC), further, deplored the artificiality of
circular maps still in use in his own day because they distort relative distances
(*Introduction to Phaenomena* 16.4.5 [Aujac]).[69]
Included on the shield is Augustus's triple triumph. Vergil's description
emphasizes the borders of the inhabited world by featuring conquered peoples
and places from each of the cardinal points of the globe (*Aen.* 8.724–728):[70]

incedunt victae longo ordine gentes,
quam variae linguis, habitu tam vestis et armis.
hic Nomadum genus et discinctos Mulciber Afros,
hic Lelegas Carasque sagittiferosque Gelonos
finxerat; Euphrates ibat iam mollior undis,
extremique hominum Morini, Rhenusque bicornis,
indomitique Dahae, et pontem indignatus Araxes.

---

69. Germaine Aujac, *Geminus, Introduction aux phénomènes* (Paris: Les Belles Lettres, 1975).
For Geminos's date, see James Evans and J. Lennart Berggren, *Geminos's Introduction to the Phenomena: A Translation and Study of a Hellenistic Survey of Astronomy* (Princeton: Princeton University Press, 2006), 17–22.

70. The triumph depicted on the shield was likely not intended to represent an actual triumph,
but it was rather a literary composite with deep symbolic resonance. Some of the races rendered on
the shield may be too obscure to agree with a historical triumph: Philip R. Hardie, *Vergil's* Aeneid:
Cosmos and Imperium (Oxford: Clarendon, 1986), 355; Östenberg, "Demonstrating the Conquest
of the World," 156, 159.

FIGURE 3.2. Aeneas's shield
as a map of the Roman
Empire, drawing by author.

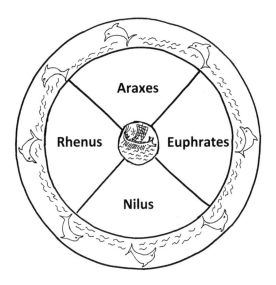

In a long line proceeded the conquered peoples as varied in language as
in custom and clothing and armor. Here Mulciber has molded the race
of Nomads and the loose-girt Afri, here the Lelegae and the Carae and
the arrow-bearing Geloni. Now gentler with its waves there flowed the
Euphrates, the Morini, the limits of men, and the two-horned Rhine, and
the unconquered Dahae, and the Araxes resenting its bridge.

Marching in the triumph is a long line of conquered (previously unconquerable?)
races who vary in their languages, dress, habit, clothes, and weapons. Vergil
highlights the heterogeneity of these conquered peoples, and this reflects the
purpose of Octavian's three-day triumph of 29 BC: "to demonstrate the conquest of the world."[71] Augustus has brought the entire world, or least its edges,
to Rome. As with early Greek maps, territory and hegemony are defined by
boundaries.

Recalling the tripartite paradigm mocked by Herodotus (see *Hist.* 4.36.2;
4.45.2),[72] Vergil has divided the external world into quadrants, four non-Roman
landmasses that envelop the Empire. Vergil arranges this triple triumph geographically by naming individual races together with their regional rivers. These
peoples are drawn from the outposts of Empire and beyond, from regions whose
mores are not Roman.

71. Östenberg, "Demonstrating the Conquest of the World," 156.
72. For Caesar's tripartite division of Gaul (*Gallic War* 1.1.1) and Pomponius Mela's tripartite
division of Spain (*Chor.* 2.87), see Irby, "Tracing the Orbis Terrarum from Tingentera."

Southward are the wandering *Nomades* of Africa (*Aen.* 8.724) and the *Afri* who are unusual because their flowing robes were worn unbelted, according to Livy (*Ab Urbe Condita* 35.11.17). They are a bellicose people who fight from horseback (in contrast with the Roman citizen infantry). Notoriously, Antony found support in Cyrene and from Bogudes, the abrogated king of eastern Mauretania (Cassius Dio, *Roman History* 48.45.2; Plutarch, *Antony* 61).[73] Thus Augustus has pacified the allies of his enemy.

From the east come the *Leleges* and the *Carians* of Asia Minor. We are meant, of course, to think of Roman defeats in Parthia under Crassus (Cassius Dio, *Roman History* 40.12–30) and Antony (Cassius Dio, *Roman History* 49.22–33). The piratical Carians are *barbari* (Vitruvius, *De Architectura* 2.18.12; Servius, *ad Aen.* 8.725), hence not Roman, and the Leleges are belligerently "arm-bearing" (*armiferi*; Ovid, *Metamorphoses* 9.645). Accompanying them are the painted (*pictos*; Vergil, *Georgics* 2.115) and quiver-bearing (*pharetratos*; Horace, *Carmina* 3.4.35) *Geloni* of Scythia, who hail from the ends of the earth (Vergil, *Georgics* 2.114–115: *extremis domitum cultoribus orbem*; Horace, *Carmina* 2.20.19: *ultimi*). Arrows are not in the typical arsenal of Rome's citizen infantry, and Rome employed allied units of archers from Syracuse (as early as 217 BC), Crete, Syria, and Anatolia (see Livy, *Ab urbe condita* 22.37.8).[74] Warpaint, furthermore, was anathema to Roman discipline.[75]

Westward are the *Morini* from northern Gaul "who gaze towards Britain" (Servius, ad loc.: *qui Britanniam spectant, proximi Oceano*), as far west as the Empire reached at the time. Their liminality as "the furthest of races" is emphasized elsewhere.[76] Vergil, moreover, subtly elicits the successful Gallic campaigns of Augustus's ancestor.

Northward, we have the Aras River, "barbarous" like the Carians (Lucan, *Pharsalia* 1.19), and the *indomiti Dahae*. According to Servius, the banks of the Aras are inhabited by Scythian nomads from the other side of the Caspian and their threatening (*minati*) Parthian neighbors who dwell beyond Roman hegemony (Seneca, *Thyestes* 603). Both are associated with Alexander's campaigns and may be included here to link Augustus's achievements to Alexander's.[77]

---

73. Duane W. Roller, *The World of Juba II and Kleopatra Selene: Royal Scholarship on Rome's African Frontier* (London: Routledge, 2003), 91–97.

74. Paul Holder, "Auxiliary Deployment in the Reign of Hadrian," *BICS* 46 (2003), 133, 135, 140.

75. Cf. Caesar, *Gallic War* 5.14.2: *vitro inficiunt, quod caeruleum efficit colorem*; Pomponius Mela, *Chor.* 3.51: *incertum ob decorem an quid aliud—vitro corpora infecti*; Martial, *Epigrams* 11.53.1: *caeruleis Britannis*; Ovid, *Amores* 2.16.39 describes them as "green" (*virdes*); Pliny, *Nat. Hist.* 22.2: *toto corpore oblitae*; Propertius, *Elegies* 2.18c.23.

76. Pomponius Mela, *Chor.* 3.23: *ad ultimos Gallicarum gentium Morinos*; Pliny, *Nat. Hist.* 19.8: *ultimique hominum existimati*. The Morini may have processed in Gaius Carrinas's triumph of 28 BC: Cassius Dio, *Roman History* 51.21.5–6; Östenberg, "Demonstrating the Conquest of the World," 159.

77. Östenberg, "Demonstrating the Conquest of the World," 161.

Significantly, the cardinal points are characterized (if not separated) by water-ways, and thus they replicate the enduring paradigm of the tripartite world that was ridiculed by Herodotus: the Euphrates at the eastern edge and Rome's eastern border (see Horace, *Carmina* 2.9.21–34), the Rhine roughly at the north-westerly edge, and the Caspian at the northeasterly extremity. The grieving Nile, defeated by Augustus, marks the south (Vergil, *Aen.* 8.711–713):[78]

contra autem magno maerentem corpore Nilum
pandentemque sinus et tota veste vocantem
caeruleum in gremium latebrosaque flumina victos

Moreover the cerulean Nile, grieving in his great body and spreading out his folds and with his entire robe calling his conquered people into his lap and into his furtive currents.

At the center of the shield is Actium, where Augustus defeated Antony and Kleopatra in 31 BC (Vergil, *Aen.* 8.675):

in medio classis aeratas, Actia bella.

In the middle the bronze-fitted fleet, the Actian fight.

In this central scene, Vergil emphasizes the conflict between east and west, foregrounding the heterogeneity of the eastern foes. Antony's allies offer "bar-baric aid" (*ope barbarica*), and their weapons are varied (*variisque armis*; *Aen.* 8.685), like the languages of the defeated peoples processing in the triumph (*variae linguis*).

Putnam observes that the placement of Actium at the shield's center shows its "importance for the poet's new cosmos."[79] Indeed, Actium was perhaps the most symbolically resonant site in Augustus's new Roman order, and Vergil was careful to include the site in his survey of the eastern Mediterranean in book three. Aeneas and his fleet had stopped there to make sacrifices, to celebrate games (in anticipation of Augustus's Actian games),[80] and to dedicate a shield

---

78. The personified Nile was displayed at Augustus's historical triumph: Propertius, *Elegies* 2.1.31–32.

79. Putnam, *Vergil's Epic Designs*, 136; Vergil, *Aen.* 8.675.

80. The games may have occurred quadrennially (Strabo, *Geogr.* 7.7.6) or quinquennially (Sue-tonius, *Life of the Divine Augustus* 18.2). For Augustus's quinquennial Actia see Robert Alan Gurval, *Actium and Augustus: The Politics and Emotions of Civil War* (Ann Arbor: University of Michigan Press, 1998), 65–67, 74–81. For quadrennial games: William M. Murray and Photios M. Petsas, *Octavian's Campsite Memorial for the Actian War* (Philadelphia: American Philosophical Society, 1989), 10.

to Apollo (*Aen.* 3.278–293). Ancient maps usually showed the mapmaker's place of origin, or some other symbolically resonant place. As in the *Ion,* many early Greek maps showed the important pan-Hellenic religious site at Delphi as the center of the world where would-be emigrants sought divine favor (*supra,* pp. 90–91). The early fourth century AD Peutinger Map, derived from Agrippa's map of the Empire, showed Rome at the center.[81] With regard to the *pax Augusta* no site had greater resonance than Actium.

The triumphal scene on the shield serves two purposes. It shows Augustus's conquest of the world, as others have already observed. And it renders, in broad strokes, a map of that new world, a powerful symbol of conquest. Victorious generals had previously manipulated the image of the vanquished *orbis terrarum* in their propaganda initiatives. Pompey had displayed a model of the world in his triumphs.[82] But we do not know if Caesar also processed in triumph with an image of the world (he did, however, commission a world map). Additionally, *Commentaries on the Gallic War* famously opens with a map of Gaul (1.1), and Caesar's account of his British campaigns, moreover, includes a narrative map of the island (5.13). In honor of Caesar's triumph, the Senate had declared a statue of the dictator standing atop the *oikoumene* (Cassius Dio, *Roman History* 43.12.2; 43.14.6).[83] Historical triumphs commonly included representations of conquered rivers with explanatory placards (Ovid, *Art of Love* 1.213–228; *Letters from Pontus* 2.1.37–44; 3.4.105–112; *Tristia* 4.2.37–46), and rivers are often poetically presented as the extremities of the Roman Empire. Ancient authors paired the Rhine and Nile, in particular, to contrast the northern and southern limits (Catullus, *Carmina* 11.8.12; Martial, *Epigrams* 4.11.7–8).[84] Nothing could better express conquest of the *orbis terrarum* than the image of the world and its conquered peoples, penned by Vergil within the triumph on Aeneas's shield.

Geography and *Romanitas*

Aeneas's shield, thus, replicates the variety of human (Roman) experience that we see in Achilles' shield where two cities—one at war, another at peace—become a full civilization. Vergil's shield is a précis of Roman history, culture, and values. And this presentation of *Romanitas* is completed by other episodes

---

81. Cf. Richard J. A. Talbert, "Peutinger Map," in *EANS,* 640 ; Richard J. A. Talbert, *Rome's World: The Peutinger Map Reconsidered* (Cambridge: Cambridge University Press, 2010).

82. Östenberg, "Demonstrating the Conquest of the World," 158.

83. Claude Nicolet, *Space, Geography, and Politics in the Early Roman Empire* (Ann Arbor: University of Michigan Press, 1991), 39–41.

84. See Östenberg, "Demonstrating the Conquest of the World," 160.

in the Augustan epic. According to Vergil's Jupiter, Roman terrestrial hegemony will extend to the celestial sphere. We recall that Anaximander's interests
included both terrestrial and celestial cartography, two trajectories which cannot
be decoupled. The same arc is evident on Achilles's doubly bifurcated shield
(two human cities; two "maps," one celestial, one terrestrial) and in the tapestries of Euripides's *Ion*.

In Vergil's *Aeneid*, two scenes emphasize the Roman geographical view of
the world. As Aeneas returns by night to camp with Arcadian allies, including
the king's son Pallas, the young prince asks the Trojan hero about the course of
the stars (10.160–162: *Pallasque sinistro adfixus lateri iam quaerit sidera, opa-
cae noctis iter*). Pallas's queries imply that Aeneas possesses such knowledge,
a knowledge that is essential for celestial navigation at night.[85] More directly,
in book six, while Aeneas surveys the parade of future Roman heroes, including
his own descendants Julius Caesar and Augustus, Anchises extends the Roman
Empire beyond the regions rendered on the shield, where liminal people dwell
just beyond Roman hegemony (e.g., the Dahae; 6.789–807). Rome will expand
eventually to India, and then to the territory of the Garamantes, a land "beyond
the stars" (6.795: *extra sidera*) and "beyond the yearly path of the sun" (6.796:
extra anni solisque vias) to where Atlas supports the sky "fitted with shining
stars" (6.797: *stellis ardentibus aptum*) and turns it "on its axis" (6.797: *axem
umero torquet*).[86] Anchises describes the apprehension of contemporary allies
and enemies. Now "agitated" (6.800: *trepida*) are the Nile and the inhabitants
of the Caspian and Maeotis. Augustus's geopolitical advancements will outstrip the mythically impressive exploits of even the divine Hercules and Dionysus. The stellar map is at best impressionistic, the stars travel along their *vias*
(or *itinera*), but their (unspecified) deployment is orderly and predictable, as in
Homer and Euripides.

By accentuating both the edges and the center, within the ecphrasis of
Aeneas's shield, Vergil draws a narrative map of the Roman (Augustan) world
whose contours come into focus as the epic progresses. The world on Aeneas's
shield has an outermost edge of dolphins frolicking in the white caps,[87] and
the interiors are fleshed out as each of the cardinal points is represented. The
conceptual map on the shield also has a physical center, with particular his-
toric and symbolic resonance, at Actium. The shield metaphorically shows that
Augustus has conquered the entire world.[88] Represented at Augustus's triumph

---

85. Williams, *The Aeneid of Vergil*, ad loc., who takes *iter* as referring strictly to Aeneas's path
and not the course of the stars. The two courses, however, overlap here, and that is the intent of
the scene.

86. Cf. Euripides, *Ion* 1147.

87. Vergil deliberately eschews explicit mention of "Ocean" in the shield ecphrasis.

88. Östenberg, "Demonstrating the Conquest of the World."

are peoples from the distant reaches of the inhabitable world, from each of the four cardinal points, and each separated by waterways that represent those cardinal points. As in Homer, Aeneas's shield consciously attempts to structure and control the world in representing the now subdued corners of the *orbis terrarum*. By traveling the *orbis terrarum*, Aeneas has given structure to it, and he has introduced Trojan-Roman control over it. Aeneas's historical counterpart, Augustus, imposed cartographical harmony and control by subduing the borders of his *orbis terrarum* (as on the shield). In the underworld we see that the Empire which is delimited on Aeneas's shield will reach beyond earthly borders. This, in fact, is Jupiter's promise to his daughter: that her son Aeneas will found an empire without limits of time or events (Vergil, *Aen.* 1.278–279: *nec metas rerum nec tempora pono*), one that will extend to the stars (1.259–260: *ad sidera caeli*).

Conclusion

Geography as represented by maps is appealing and powerfully symbolic, and its interpretation can be deeply personal or profoundly political. The Roman epigrammist Martial (40–104 AD), for instance, catalogues the hometowns of Roman authorial giants from Catullus to Valerius Flaccus, crafting a landscape that speaks to a broad compass of literary culture (*Epigrams* 1.61). Martial thus creates an elegiac map, by listing toponyms, the distinguishing climactic and topographic features, and the famous sons of those literary cities. The Nile is rain-bringing (*imbrifer*), Corduba is eloquent (*facunda*), Gades rejoices (*gaudent*), and Bilbilis will boast (*gloriabitur*). Town and poet become one. Geographical expertise flowed deeply, and Martial's readers recognized Catullus by his hometown of Verona, Vergil (obliquely noted by his cognomen Maro) by his hometown of Mantua, and Ovid by the Paeligni (the peoples who inhabited his native region of Sulmo in central Italy). In this twelve line epigram, Martial cites twelve authors, six of whom are from Spain: the two Senecas, Lucan, Canius and Decianus, both otherwise unknown, and Martial himself. Thus the epigrammist's home province is foregrounded, and the poet writes himself into a robust literary tradition, "putting himself on the map," so to speak. We note that none of Martial's authors hailed from Rome. The epigram, furthermore, is emphatically framed by two resonant toponyms: Verona, the hometown of Martial's literary model, and Bilbilis, his own birthplace, respectively the first and last words of the piece.

Geography and cartography serve as powerful tools whereby authors impose order and geopolitical control over their worlds. Cartography furthermore assumes many forms whether it be the imaginary landscape sketched by

Apollonios of Rhodes in the *Argonautica* or the composite political maps rendered in Euripides's *Ion* or on Aeneas's shield. Ulysses scratched the walls and streets of Troy into the wet sand of Calypso's beach,[89] and Julius Caesar, a man who deeply appreciated the value of geography for understanding and controlling provincial territory, drafted graphic battle plans (e.g., Caesar, *Gallic War* 5.1.1). That Rome appears dead center in Pomponius Mela's geography was a political and diplomatic decision, but it was a deliberate and highly idiosyncratic choice to position Spain and to place Tingentera just off-center.[90] Anaximander likely spiraled his own *oikoumene* around Miletos. For Euripides, Delphi was the center of Apollo's world. Vergil, consequently, positioned Actium at the center of Augustus's world on Aeneas's shield. The literary landscape is thus created from a knowledge and manipulation of scientific geography, which in turn is employed and interpreted to underscore the cultural resonance of place, boundaries, identity, and hegemony.

89. In *The Art of Love* 2.125–142, Ovid recounts a charming vignette of a lazy seaside afternoon spent between Odysseus and Calypso. Calypso asks, again and again, that her captive lover recount his adventures in the Trojan War. Again and again, Odysseus "finds fresh words for the same old tale." As they stroll along the beach, Calypso asks about one battle or another, while Odysseus sketches out the scenes for her in the wet sand. "This, he says, is Troy (he made walls on the shore). This is the Simois. Let's consider this spot my camp."

90. See further, Irby, "Tracing the Orbis Terrarum from Tingentera."

CHAPTER 4

Tracing the *Orbis terrarum* from Tingentera

Georgia L. Irby

ONE OF SEVERAL IBERIAN INTELLECTUALS at the Roman court in the mid-first century AD, Pomponius Mela was the first systematic geographer writing in Latin whose work survives intact. He composed under Claudius and may have intended his *Chorography* to celebrate that emperor's British triumph (3.49–52).[1] Despite scholarly precedent, Mela took his own route around the Mediterranean. While starting at the traditional spot, the Pillars of Hercules, his maverick decision to work counter-clockwise around "Our Sea" (*Mare Nostrum*) nonetheless allows him to place his hometown, Tingentera, an obscure Phoenician settlement near the Strait of Gibraltar, roughly at the center of the treatise and of his cosmos. Just as Apollo's sanctuary at Delphi stood at the center of the early Greek geomythical world, so the Pillars of Hercules, which structure both Mela's worldview and his text, serve as his personal *omphalos*. Had Mela followed the conventional clockwise path, Tingentera would have been quickly described, passed over, and forgotten early in the opus. Instead, Mela deliberately brings his hometown just off-center.

Mela's Iberian origins linger in the background, and his work reflects complex tensions between these two cultural identities, Iberian and Roman. On the one hand, Mela features Rome at the physical center of the *Chorography* (2.60),

Author Note: Earlier versions of this paper were delivered at the *Scientists and Professionals in the Ancient World Conference,* School of Classics, University of St Andrews, Scotland, 7 September 2009, and the *Classical Association of the Middle West and South,* Grand Rapids, MI, 8 April 2011. With sincere thanks to Duane W. Roller for his insightful critique on this work from the inception.

1. *De Chorographia* seems to have been composed after the invasion of AD 43 but before the triumph in February 44. See also Cassius Dio, *Roman History* 60.23.1. While Mela refers to Claudius's invasion of Britain explicitly, he mentions the emperor only indirectly: "behold, the greatest of emperors (*principum maximus*) is opening up Britain which had been shut for so long (*clausam*)." In a touch of whimsy, Mela crafts a transparent and perhaps not too clever pun on Claudius's name, widely recognized as resembling the verb meaning "to close" (*claudo*). Some twelve or so years later, Seneca the Younger would make a similar pun in the *Apocolocyntosis*: when the recently deceased Claudius is knocking on the door to Olympus, trying to persuade Hercules to allow him his proper place amongst the deified gods, Hercules responds: "It is no wonder that you burst through the doors of the *curia*: nothing is closed to you (*nihil tibi clausi est*)" (*Apocolocyntosis* 8).

but his trite summary of the imperial capital, barely described in a survey of Italian cities, contrasts sharply with the hyperbole with which he treats the Iberian peninsula. In addition, he foregrounds Hercules, a mythological hero whose territory included the extreme western edge of the Mediterranean and whose cult was particularly vigorous in southern Iberia. Furthermore, with his inclusion of *paradoxa*, Mela exploits the aesthetic and intellectual tastes of the Roman Silver Age, a time when bizarre hybridism was popular. Other scholars have been quick to point out errors, but it is not our aim to "correct" Mela's mistakes. Although Mela may declare his cultural *Romanitas* in many ways, his presentation of geography is a complex harmony of his Roman, Silver Age, and Tingenteran outlook.

Geographical Writing at Rome Before Mela

Let us first consider the history of geographical writing at Rome, starting with Julius Caesar who, in 46 BC, commissioned four Greek surveyors from Alexandria (one for each of the cardinal directions) to chart the lands around the Mediterranean as the foundation of a narrative map of the entire (Roman) world. Caesar's interest in geography and cartography long preceded his dictatorship. As a successful provincial governor and general in the field, he was deeply aware of the utility of topography and geography in managing affairs, anticipating Strabo's advice that effective leaders should know "the size of a country, lay of the land, peculiarities of sky and soil" as aids to "pitch(ing) camp, set(ting) an ambush, and march(ing) in unfamiliar territory."[2] Caesar's *Commentaries on the Gallic War* famously open with a literary map of the three culturally and linguistically distinct provinces of Gaul (1.1; we return to the general's map below). His account of his British campaigns also includes a narrative map that offers, in addition to the resources and peoples of the island, practical data that would be of use to a governor or military commander: shape and dimensions, neighboring bodies of land, and the surprising behavior of the midwinter sun where darkness endures for thirty days (5.13).[3] Ever the pragmatic Roman administrator, Caesar makes note of rivers which provide handy landmarks,[4] mountain

2. Strabo, *Geogr.* 1.1.1 on the utility of geography.

3. The locals were unable to explain the phenomenon, but Caesar's own measurements (presumably with a *klepsydra*) showed that British summertime nights were shorter than on the continent, owing to Britain's higher latitude.

4. For example, the deep and broad Rhine (Caesar, *Gallic War* 1.2), the slow-flowing Saône (1.12); the river Doubs, which, moat-like, encircles Vesontio, a town of the Sequani (1.38); the river Aisne, which marks the extent of the Remi territory (2.5); the river Rhone, which demarcates the territory of the Nantuantes, Veragri, and Seduni (3.1); the river Loire, which debouches into the Atlantic (3.9); the Tamesis, a tidal river fordable only in one spot where a nasty native defense system was

ranges,[5] the ease and length of marching routes,[6] distances from the enemy,[7] supply lines, and battlefield topography.[8] He takes particular care in describing the extent and marching conditions of the Hercynian forest (6.24–26), as well as the tactical challenges of the topography at Alesia (7.69) and Dyrrachium (3.44–46), among other places.

Caesar's world-map project was never completed. But his heir, his great-nephew Octavian (Augustus, as he came to be known), eventually tasked the talented Marcus Vipsanius Agrippa with continuing the dictator's work. Agrippa's map was completed posthumously by Augustus, and publically rendered in marble on the walls of the Porticus Vipsania, near the Ara Pacis. Over the years the map may have been updated to document the Roman conquests of Britain, Dacia, and Mesopotamia. Intended in part to aid in vanquishing and taxing mapped areas,[9] it was a powerful symbol of imperial hegemony that represented the entire inhabited world as known to the Romans in the 20s BC,[10] and it was likely erected to stress that Rome (that is to say, Augustus) had unified the entire world under the *Pax Augusta.* Unfortunately, the original artifact is lost. But it was copied and reinterpreted, and it likely inspired the Peutinger Map.[11] The Augustan map was probably accompanied by a commentary, cited thirty-six times by Pliny the Elder who, nonetheless, provided no extracts. His interest was simply in Agrippa's distance measurements, and he questioned Agrippa's authority only once.[12]

For Caesar and Agrippa, geography was a practical and dispassionate tool of efficient administration. Caesar's geographical outlook can be pieced together from data scattered throughout his *Commentaries.* Agrippa's principles of geography were likely in parallel with Caesar's.[13] But geography was treated more

---

presumably set in place (5.18); an unnamed steep-banked river that was difficult to ford and thus a welcome barricade against the enemy (6.7).

5. E.g., the Jura Mountains, which separate the Helvetii from the Germans (Caesar, *Gallic War* 1.2); the Alps, whose crossings were dangerous and regulated with tolls (3.1); the Vosges mountain range, whence the Meuse River flows into the territory of the Lingones (4.10).

6. Nineteen Roman miles from lake Lemannus to the Jura mountains (Caesar, *Gallic War* 1.8); the Hercynian forest can be crossed in nine days by a lightly armed man on foot (6.25).

7. As where Caesar maintained a distance of no more than five Roman miles between the enemy rearguard and his vanguard (Caesar, *Gallic War* 1.15); where the enemy camp is just over seven miles from the Roman camp (1.21).

8. Caesar, *Gallic War* 1.26; 2.9; 2.23.

9. Pliny the Elder, *Nat. Hist.* 3.17; Tracey Elizabeth Rihll, *Greek Science* (Oxford: Oxford University Press, 1999), 83–84.

10. O. A. W. Dilke, *Greek and Roman Maps* (Ithaca, NY: Cornell University Press, 1985), 41–53.

11. Richard J. A. Talbert, *Rome's World: The Peutinger Map Reconsidered* (Cambridge: Cambridge University Press, 2010).

12. Pliny the Elder, *Nat. Hist.* 4.91, asserts that any measurements of remote parts of the world— like the Borysthenes (Dnieper), for which Agrippa provides dimensions—were at best uncertain.

13. Dilke, *Maps,* 39–54; G. L. Irby-Massie, "M. Vipsanius Agrippa," *EANS,* 830.

holistically and synthetically by subsequent Latin authors, including Pliny the Elder, who devotes four of his thirty-seven book encyclopedia to the topic, and Pomponius Mela.

Mela was hardly an aspiring scientific geographer: he lacked the technical preparation.[14] Even the casual reader will notice that Mela usually omits distances.[15] But our author is curious about the world, including its size,[16] how the earth has changed over time,[17] and the length of daylight in different parts of the world.[18] Even if Mela had known Strabo's work, he would likely have felt no compulsion to rewrite the *Geographika* in Latin.[19] Mela offers something new, a Greek-style *periplus* ("coasting guide") with a Silver Age patina, a Roman conception of the geography of the inhabited world in Latin for an educated audience. It is, however, difficult to gauge Mela's impact: Pliny includes Mela among his Latin geographical authorities for books three through six but never cites him by name in the text. Mela, nonetheless, was consulted by Solinus

14. Frank E. Romer, ed. and trans., *Pomponius Mela's Description of the World* (Ann Arbor: University of Michigan Press, 1998), 7n13, 25, and 34n4, suggests that Mela did not understand the cartographic ramifications of a spherical earth. For example, at 1.54 in describing those who dwell opposite us (*nobis*) in the south (*a meridie*), he employs Antichthones instead of the usual Antipodes. The two terms are, nonetheless, "functionally interchangeable:" James S. Romm, *The Edges of the Earth in Roman Thought* (Princeton: Princeton University Press, 1992), 131. At any rate, Mela is inconsistent. He describes the Antichthones as unexplored and unknown (*Chor.* 1.4), but later expresses skepticism about the region's very existence (1.54). For the Antipodes and the spherical earth: Aristotle, *On the Heavens* 308a20; Cicero, *Academia* 2.123. Like Mela before him, Pliny questions the existence of the Antipodes (Book One, in his table of contents for 2.65–66; cf. 2.178; 4.90), an issue which he tackles in terms of theoretical astronomy. The "other part" (*altera pars*) of the earth is also unknown (*ignotae gentes*) in Manilius, *Astron.* 1.377–378.

15. Dimensions cited by Mela: Mediterranean Sea (*Chor.* 1.6); Gulf of Syrtis (1.35); Black Sea (2.4); Borysthenes River (2.6); the length of the Indian coastline (3.61); the breadth of the Greek mainland (2.37); and the area occupied by the pyramids (1.55).

16. India is so large that in some places the celestial pole is not visible (Mela, *Chor.* 3.61).

17. Mela pauses to describe the creation of Strait of Hercules (*Chor.* 1.27). He explains that Pharos (2.104) and Tyre (1.66) were once islands, and Sicily was once attached to Italy's toe (2.115). Although skeptical, he punctiliously reports inland deposits of marine remains (1.32; cf., Xenophanes, *TEGP* 59; Herodotus, *Hist.* 2.12; Strabo, *Geogr.* 1.3.3–4). Egyptian civilization is so ancient that the stars have changed their courses four times and the sun has set twice already where it now rises (1.59), a clear reference to the "Great Year" whose existence had been theorized by Plato (*Timaeus* 39d) and proved by Hipparchos of Nicea's (ca. 130 BC) discovery of the precession of the equinoxes (Ptolemy, *Almagest* 3.1, 7.1–2). See Otto Neugebauer, *Astronomy and History: Selected Essays* (New York: Springer, 1983), 320–25.

18. Mela twice mentions the extremely long days that occur in the higher latitudes during the summer: in Thule, far to the north, there is no night during summer solstice (3.57); the Hyperboreans, who dwell under the pole star, experience six months of daylight and six months of night (3.36).

19. The exhaustive Pliny the Elder makes no mention of Strabo, nor does Marinos of Tyre or Plutarch, who both would have found the work a valuable source. Plutarch did cite Strabo's *Histories* (see *Sulla* 26.3; *Lucullus* 28.7; *Caesar* 63.2): Duane W. Roller, *The Geography of Strabo: An English Translation with Introduction and Notes* (Cambridge: Cambridge University Press, 2014), 27 and n78.

(third century), Juvenal's scholiast (fourth century), Vergil's commentator Servius (fourth/fifth century), Martin Capella (d. AD 397), and Isidore of Seville (d. AD 636).[20] Mela's influence may have been tenuous, but the *Chorography* endured.

Mela's Sources

Mela's *Nachleben* seems commensurate with his own use of references. He explicitly names only three authorities plus the "physicists."[21] Homer was Mela's source for the one hundred gates of Thebes[22] and for the fact that Pharos was once a full day's sail from Alexandria.[23] Mela also cites Hanno in a selective précis of the Carthaginian leader's early fifth century BC expedition beyond the Pillars of Hercules (3.90–93). Mela finds interesting Hanno's report of hirsute, savage women (Hanno's "gorillae") who inhabited an island without men. Here Mela contradicts the Carthaginian who claimed instead to have been unable to capture any of the males (on Hanno's report, the females greatly outnumbered the males).[24] Hanno also reported that "Gorilla" island lies in a lake embedded within an island in a larger lake. For Mela, there is no further nesting of Gorilla Island. A large bend of the shore encloses the sizable island.[25] These female brutes gave birth parthenogenetically, and they were often unable to be restrained by chains (details not in Hanno). Mela seems to be conflating Hanno's "Gorillae" with the Amazons, the legendary Scythian women who rejected their menfolk.[26] Claiming a preference for empirical evidence, Mela believes that

20. Mary Ella Willham, "Mela, Pomponius," in *Catalogus Translationum et Commentariorum: Mediaeval and Renaissance Latin Translations and Commentaries, Annotated Lists and Guides*, ed. Virginia Brown (Washington, DC: Catholic University of America Press, 2011), 9:258.

21. Alain Silberman, *Pomponius Mela: Chorographie* (Paris: Les Belles Lettres, 1988), xxx–xxxvi, explores more deeply Mela's likely sources for many sites.

22. Mela, *Chor.* 1.60: according to other authors Thebes has one hundred palaces.

23. Mela, *Chor.* 2.104. In Mela's day a mole connected the island to the mainland. See Homer, *Od.* 4.354–357. Strabo, *Geogr.* 1.2.23 defends Homer's "ignorance" against the unreasonable reproaches of geographic writers. Many authors had noted the interplay between earth and water, whereby land is restored by siltation or lost by erosion: Herodotus, *Hist.* 2.5; Aristotle, *Meteorology* 351a19–352b16; Strabo, *Geogr.* 12.2.4.

24. We cannot be certain if Mela consulted a copy of the Punic inscription (he may have possessed only a rudimentary reading knowledge of the Punic language), or if instead he reviewed (and misinterpreted) Juba II's commentary of Hanno's voyage. For Juba's commentary, see Duane W. Roller, *The World of Juba II and Kleopatra Selene: Royal Scholarship on Rome's African Frontier* (London: Routledge, 2003), 187–90.

25. Pliny's Isle of the Gorgons (*Nat. Hist.* 6.200). Mela would have found irresistible Hanno's convoluted image of lakes nested in islands nested in lakes.

26. Mela is aware of the Amazons, citing their territory as a geographical point of reference several times: 1.12; 1.13; 1.109; 3.39. He credits the Amazons with the dedication of the temple of

artifacts, the three hides which Hanno conveyed to Carthage—just the sort of curiosity that appealed to Silver Age tastes—gave credence to the account (3.90; 3.93).[27] Mela also cites by name Cornelius Nepos as the authority for Eudoxus of Cyzicus's successful circumnavigation of Africa from the Arabian Gulf to Gades (118/116 BC; 3.90; see also Strabo, *Geogr.* 2.3.4–5). Although the theory had been promoted by Homer and unnamed natural philosophers (*physici*), Mela finds Nepos's account of the circumambient Ocean as all the more reliable because Nepos was the more recent authority (3.45).[28]

Elsewhere, Mela vaguely differentiates between indeterminate Romans and Greeks, and he is occasionally critical of non-Roman sources. For example, Mela employs etymological evidence to show that Greek authors were in error in their fantastical claims that Liber was gestated in Jupiter's thigh (*meros*), when the god, according to Mela's understanding of local tradition, was born on Meros, a mountain sacred to Jupiter, above the legendary city of Nysa in India (3.66. See also Strabo, *Geogr.* 15.1.8). Mela is also astounded to find that even "our" unnamed authors (*nostris auctoribus*), alongside anonymous Greek writers, were perpetuating an absurd story about a fish pulled out of the earth near the Salsula spring (listed in the Antonine Itinerary as on the road from Narbo to the Pyrenees; 2.83: *Grais nostrisque etiam auctoribus*).[29] Mela's use of "even"

Artemis at Ephesus (1.88). And he alludes to their wandering: they were expelled from Aeolis when Cyme was their queen (1.90); they once encamped on the plain of Lycastos at a site now known as Amazonius (1.105); and the Gynaecocratumenoe ("the folk ruled by women") at one time inhabited "kingdoms of the Amazons" which they had established along the banks of the Tanaïs on lands good for pasturing but little else (1.116). Although not strictly calling them Amazons, Mela also describes the women of the (Thracian) Ixamatae who dwell near the Maeotis (1.114). They engage in the same activities as men, battling on horseback and attacking the enemy with lariats (instead of swords), but these women do marry, unlike the Amazons of myth. Mela also makes note of the Sarmatian women whose right breasts are cauterized at birth, in order to facilitate archery and military service (3.34–35). Those women archers who fail to kill the enemy are punished with perpetual virginity.

27. See Hanno, *Periplus* 17–18. Hanno's "gorillae" may have been chimpanzees or baboons. Hanno dedicated the hides in the temple of Juno/Tanit at Carthage where they remained on display until 146 BC when Carthage was razed: Pliny the Elder, *Nat. Hist.* 6.200. Pliny reports that Hanno brought back only two hides.

28. Pliny the Elder, likewise, gives greater authority to recent, contemporary, or first-hand geographical reports. Pliny's account of the lands beyond the Atlas mountain range in northwestern Africa was informed and vivified by his contemporary, Suetonius Paulinus, the first Roman commander to cross the range (Paulinus was propraetor of Mauretania in 42). Pliny, however, restricts his own eye-witness accounts to technical achievements, describing in painstaking detail, for example, the fertility of a plain in Byzacium (Africa: *vidimus* 17.41) and a method of mining in Iberia (*fecimus*: 33.70–73).

29. Cf. Strabo, *Geogr.* 4.1.6, who supplies more detail. Subterranean rivers are not entirely uncommon, especially in Greece, and they were a source of curiosity for natural historians (Aristotle, *Meteorology* 349b27–351a6; Seneca the Younger, *Natural Questions* 3.7–10). For fishes adapted to subterranean and cave environments, Aldemaro Romero and Kelly M. Paulson, "It's a

(*etiam*) highlights cultural as well as linguistic prejudices. We might believe that the Greeks, ignorant about the coasts of Gallia Narbonensis and Hispania Tarraconensis (despite extensive settlement in the seventh through fifth centuries BC), would perpetuate such a falsehood, but we would expect "our authors" to be better informed. Perhaps here Mela has in mind local (not necessarily Roman) authorities. He also invokes multiple unspecified sources to account for conflicting data, thus allowing his readers to come to their own conclusions. The inhabitants of Caria, for instance, might be indigenous, as in some sources, or Pelasgian, as others claim, or even Cretan, as yet others suggest (1.83). Our Iberian author aims to appear scholarly without burdening the reader with yet another list of sources which would add to the tedium of a topic that consists largely in "the names of races and locales and their convoluted deployment," a fact for which the author apologizes in his opening words (1.1).

It is however impossible to reconstruct Mela's sources, and Silberman suggests that our writer likely consulted most of his authorities indirectly (Herodotus, Eratosthenes, Polybius, among others).[30] Herodotus's influence is particularly evident in ethnographies of Egyptian (1.57–60) and Scythian (2.9–15) peoples, where our Iberian author selectively distills the historian's rich details. Mela's perfunctory account of the Nile's annual flooding may also have been condensed from Herodotus (1.53; Herodotus, *Hist.* 2.20–25).[31] The chorographer perhaps drew from Caesar on the Druids,[32] and almost certainly for his ethnography of the Germans which aligns closely with the general's description (3.26–28; Caesar, *Gallic War* 6.21–24).[33] Finally, Mela's description of tidal patterns corresponds with earlier hypotheses and contemporary Stoic theory: an animate cosmos, Plato's subterranean caves, and the lunitidal theory refined by Poseidonios who, incidentally, had visited Gades to study the tides there (3.2).[34]

Wonderful Hypogean Life: A Guide to the Troglomorphic Fishes of the World," in *The Biology of Hypogean Fishes*, ed. Aldemaro Romero (Dordrecht: Kluwer, 2001), 13–41.

30. Silberman, *Mela*, xxxiii–xxxiv. Consider, for example, Mela's substitution of Garamantes (1.23) for Herodotus's Gamphasantes (4.183–184).

31. Mela's causes of the annual Nile flood include snowmelt, proximity of the sun to the earth in winter, and the etesian winds.

32. Like Caesar (*Gallic War* 6.13), Mela (*Chor.* 3.18–19) emphasizes the secrecy of Druidic teachings, but he omits Caesar's mandate against committing Druidical wisdom to writing. Both authors also specify that Druidical training lasts up to twenty years and that the Druids teach the transmigration of souls. For Caesar, transmigration explains the fearlessness of the Celts in battle; for Mela, the doctrine sheds light on the practice of grave goods that are "suitable for the living."

33. Both authors remark on the courage of the Germans, their constant warfare, and how the Germans employ exercise and exposure to cold in order to strengthen their bodies.

34. Silberman, *Mela*, ad loc. Animate cosmos: Poseidonios f23; Cicero, *Nature of the Gods* book 2; Pliny the Elder, *Nat. Hist.* 2.98. Tides caused by the moon: Poseidonios f217.

The Organization of Mela's *De Chorographia*

It was the usual practice of Greek geographical authors, e.g., Hekataios, Strabo, and Ptolemy, that descriptions of the inhabited world began at the Pillars of Hercules (Strait of Gibraltar), proceeding clockwise (Europe, Asia, Africa).[35] Finding risible the circumambient Ocean that shaped the worldviews of his predecessors, Herodotus rejected geographical theory altogether, but his treatment of the landmasses was roughly clockwise: Persia, lands to the northeast of Persia, Libya, Europe (*Hist.* 4.36–45). Pliny rebuffed a seamlessly contiguous treatment of the world (Western Europe, Eastern Europe, Black Sea, Africa, Middle East, Asia, and India).

While Mela's chorographical organization may be unique, other geographical authors also eschewed Hekataios's archetype. Eratosthenes proceeded from east to west, perhaps in response to contemporary interests where India, newly opened up to the Greek-speaking world, was a focus of intense curiosity.[36] We see yet another geographical model in the *Astronomica* of Marcus Manilius (flourished 10–30 AD) who might have immigrated to Rome, as Mela had done.[37] Manilius may even have inspired Mela in the design of his *Chorography*.

A Roman astrological poet active in Rome just before Mela, Manilius includes a geographical ecphrasis that veered from the Greek geographical exemplar (*Astron.* 4.585–710).[38] After establishing the four cardinal points according to the quarters of the winds (north, east, south, west), Manilius guides his readers around the Mediterranean along four connected tracks.

First, he follows the sun (clockwise) through the Mediterranean, starting at Numidia, progressing then to Libya, Carthage, near the Syrtes (whose billows flow back into the Nile), Iberia, Gaul, Italy, and Sicily. From the dangerous strait between Sicily and Italy (where dwelt Scylla's greedy dogs and Charybdis's greedy maw), Manilius continues his clockwise *periplus* with a focus on the waterways as they change from one into the next: the Ionian Sea flows into the

---

35. Duane W. Roller, *Eratosthenes' Geography* (Princeton: Princeton University Press, 2010), 24; Romer, *Mela*, 9; Jacoby, "Ephoros" *FGrHist* 70.30 ad loc.

36. Roller, *Eratosthenes' Geography*, 24.

37. Manilius tells us little about himself and is cited by no other Roman author: Katharina Volk, *Manilius and His Intellectual Background (Oxford: Oxford University Press,* 2009), 1. Richard Bentley, M. Manilii Astronomicon (London: Vaillant, 1739), 19–20, who viewed Manilius's style as "peculiar," considered the poet Asiatic. Manilius's Latin is idiomatic and "correct," however sententious and elliptical it might be. Manilius was inspired by Vergil, and he sought to refute Lucretius, with whose work he was intimately familiar. But his work also reveals a deep knowledge of Greek astrological texts and Greek literary authors, both major (Homer, Euripides) and minor (including Longinus's De sublimitate): Robinson Ellis, "The Literary Relations of 'Longinus' and Manilius," CR (1899), 294; Volk, Manilius, 202n55).

38. See Romer, *Mela*, 9n17.

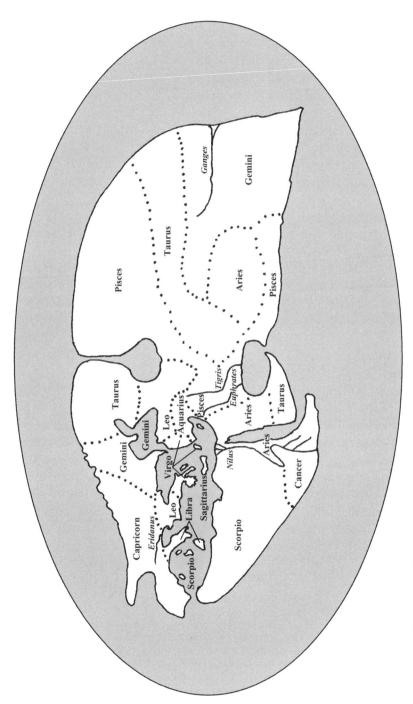

FIGURE 4.1. Tracing Manilius's cosmos, drawing by author.

Adriatic, which in its own turn completes the circuit of Italy, drinks the waters of the Eridanus, and restrains Illyria from war. Next, the sea hastens around the Peloponnese, flows past Thessaly and Achaea, and then into the Hellespont, a "reluctant" passage (*invitum*: 616). The waters subsequently glide into the Propontis and Euxine Seas, and the Maeotis (Sea of Azov), refluxing back through the Hellespont into the Icarian (southern Aegean) and Aegean seas, where Asia is on the left. From there they pass Cilicia and Syria on their way to the Arabian Gulf. Manilius's *periplus* completes its circuit in Egypt, "coming to an end again with the banks of the Nile" (4.601).

Second, Manilius traces the route of the major islands from east to west, again following the sun (Crete, the Cyclades, Delos, Rhodes, Tenedos, Corsica, Sardinia, and the Balearics off the eastern coast of Iberia).

Third, Manilius leads his reader eastward again with his treatment of Oceanic incursions into the Mediterranean (Pillars of Hercules, Hellespont, Persian Gulf, Arabian Gulf).[39]

Finally, Manilius surveys the major landmasses and their marvels in a counter-clockwise fashion: Libya, replete with disease as well as dangerous, hideous creatures; Asia, "rich in everything"; India, "too vast to be known"; Parthia and Scythia; the European lands—Greece, Illyria, Thrace, Germany, Gaul, Iberia; and finally Italy, the capital of the world. Manilius neatly returns to astrology with an exegesis of the celestial signs that rule specific parts of the world.

No doubt Manilius intended to replicate his account of the constellations in his terrestrial map: he treats the zodiac first, the northern stars, and finally the southern stars. The progression of Manilius's *mappamundi* is orderly, efficient, and logical. The reader can trace a continuous clockwise journey along four stylized paths: around the Mediterranean (from Numidia circling around back to Egypt), from eastern islands (Cyprus, "north" of Egypt) westward to the Balearics beyond the Pillars of Hercules, back eastward through the Pillars as the straits and gulfs are surveyed, and finally a counter-clockwise spiral from the Arabian Gulf/Africa to Asia, to Europe, before the journey is completed in Italy. Manilius's starting point (Numidia) may be politically charged: Juba II, a Numidian royal, had been raised at Rome and established by Augustus as an allied client-king in Mauretania. Augustus also arranged for Juba's marriage to Kleopatra Selene (daughter of Augustus's rival Marcus Antonius and

---

39. Manilius here perpetuates a common error: that the Euxine flowed into the Caspian which in turn flowed into Ocean. In the 280s BC, Patrokles, who traveled there before venturing into India, erroneously postulated that the Caspian was linked with the Ocean, and his (lost) account probably informed Eratosthenes's view that the Caspian was an inlet of the external Ocean (f110; Pliny the Elder, *Nat. Hist.* 6.58; Roller, *Eratosthenes' Geography*, 19, 206). Only Herodotus (*Hist.* 1.202.4) and Aristotle (*Meteorology* 2.1 [354a4]) recognized the Caspian as a landlocked lake.

Kleopatra VII).[40] Manilius's passing reference to the Nile River naturally elicits Augustus's (not-so-recent) defeat of Kleopatra in 31 BC (*Astron.* 4.601).[41] That Manilius would end the *periplus* at Rome, the "mistress of the world" who has been joined with heaven, is both deliberate and emphatic. Interestingly, Manilius describes the earth as "swimming," thus evoking Thales's floating disc of land (4.595: *natat tellus*; cf. Thales, *TEGP* 18–20), and he embraces the theory of the circumambient Ocean, first noted by Homer (4.596: *cingentibus medium liquidis amplexibus orbem.* See Homer, *Il.* 18.607).[42]

Likewise, Mela traces an orderly *mappamundi* in several steps. He begins with an overview of the world: Asia, Europe, and Africa. He turns to separate regions (and settlements) in more detail: Africa, Asia, and Europe. He then proceeds to explore the Mediterranean islands, following the sun from east to west, starting in the Maeotis, "whence it seems easiest to begin." Next, he describes the northern outer coasts (reminding us that he treated these from Baetica "all the way here" [Scythia]): Iberia, Gaul, Germany, Sarmatia, Scythia. Finally, returning back to Gades, the outer islands and coastal areas: Britain, Ireland, Orkneys, Thule, Scandinavia, Scythia, India, Persian Gulf, the east coast of Africa, the west coast of Africa, ending where he began, at the Pillars of Hercules near Tingentera, his hometown.

We note several comparanda. In both texts, the reader's journey is a multi-staged progression from one locale to the next.[43] Each author treats the larger landmasses, major Mediterranean islands (both accounts follow the sun), and then the space beyond the Mediterranean as it relates either to *Mare Nostrum* or the landmasses (straits/coastal lands). Each writer closes his work with a symbolically resonant location (Rome/Pillars of Hercules). Mela's course, which is efficient, deliberate, and personal, reflects the fact that, under the Romans, geography has become much more complex and nuanced. The simple Greek clockwise paradigm no longer suffices. Mela's explication of the peoples and places of the Mediterranean mirrors his own view of geography. Chorography is *perplexa*, "thoroughly folded up" (1.1), and so is the evolution of Mela's *periplus* as he traces the world according to its folds.

Both authors begin with a compass: Manilius's is based on the quarters of the winds, Mela's follows the risings and settings of the sun. These compasses

---

40. Roller, *The World of Juba II*, 84–90.

41. At length Manilius praises Augustus's birth sign, the Capricorn, and addresses the *princeps* directly (*Astron.* 2.187, 366, 417, 553, 563).

42. Herodotus (*Hist.* 4.8) and Hipparchos of Nicea argued against model of the circumambient ocean on the lack of empirical evidence, but the paradigm endured, finding support in Aristotle (*Metaphysics* 354b), Eratosthenes (f39), Poseidonios (f214), and Strabo (*Geogr.* 1.1.8–9).

43. Romer, *Mela*, 13–15, observes that Mela acts like a tour guide, taking the reader along with him on a journey that maps his own interests.

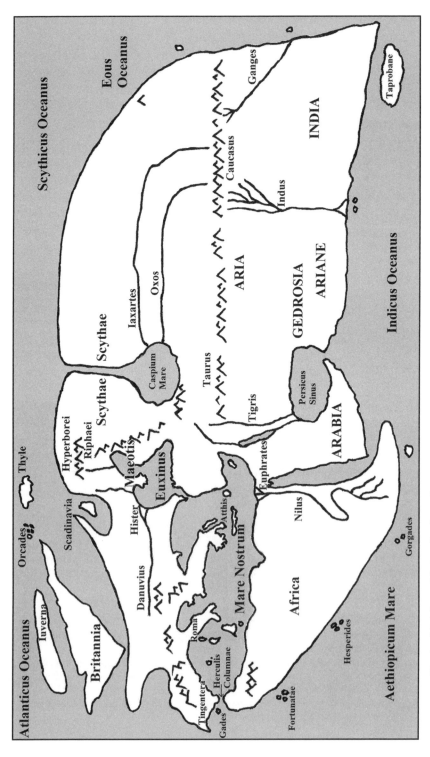

FIGURE 4.2. Tracing Mela's world, drawing by author.

derive from the geographical overview of the *oikoumene* first presented by the historian Ephoros (flourished 360–330 BC) who organized his earth as a flat rectangle and whose cardinal limits were cited according to the winds, represented by four liminal peoples: the Scythians (north), Indians (east), Ethiopians (south), and Celts (west).[44] Krates of Mallos (ca. 170–120 BC) further developed the view of a world in quadrants according to the cardinal points on his terrestrial globe, which was displayed in Pergamon about 150 BC. Showing two intersecting belts of Ocean which separated four symmetrical landmasses, the globe diverged starkly from previous maps.[45] Mela here has acknowledged but reinterpreted a cartographic convention by translating the quarters of the winds into the cardinal points of the sun, thus writing himself into the genre while simultaneously making a unique contribution.

More broadly, whereas Manilius treats the celestial ambit, Mela describes the terrestrial sphere. In his proemium, Manilius emphasizes his originality, disingenuously claiming to be the first (*primus*) poet to describe the heavenly array (*Astron.* 1.4).[46] Characterizing his work as the pursuit of "time-consuming" (rather than pleasant) material, Mela underscores the complexity of his project (*perplexi*). The topic, Mela claims, does not lend itself to eloquence because of the complex, folded-up (*perplexo*) arrangement of peoples and places in the world. Manilius, likewise, imagines spending his life (*vivere*) "touring the boundless skies, learning about the constellations and the contrary (retrograde) motions of the planets" (*Astron.* 1.13–15). His work, studying the heavens, is made possible only with the leisure afforded in times of peace (1.13: *hoc*

---

44. Strabo, *Geogr.* 1.2.28. See also Silberman, *Mela*, ad loc.

45. Strabo, *Geogr.* 2.5.10, approves the globe as a more realistic rendering of the world since it, like the earth, is spherical. Krates thus applies geometry to his interpretation of Homer, demonstrating that maps could be intrinsically valuable in academic (literary) debate. Cicero incorporates Krates's vision into the Dream of Scipio (*Republic* 6.20–21). Macrobius (*Commentary* 5.31–36) augments Krates's theory which would come to influence Medieval cartography: Dilke, *Maps*, 37.

46. The divine nature of stars and constellations was commonly espoused, stellar progressions were firmly rooted in the Greek intellectual landscape, and descriptive astrology became a legitimate topic for literary treatment. Aratus's *Phaenomena* was one of the most widely read poems in antiquity, inspiring a commentary (Hipparchos), several translations into Latin (by Varro, Cicero, Germanicus and Avienus), and overt praise from Callimachus (*Epigram* 27.3–4 [Wilamowitz]): Peter Bing, "Aratus and his Audiences," *MD* 31 (1993), 99–109; Douglas A. Kidd, *Aratus, Phaenomena: Edited with Introduction, Translation and Commentary* (Cambridge: Cambridge University Press, 1997); Emma Gee, *Ovid, Aratus, and Augustus* (Cambridge: Cambridge University Press, 2000). The *Phaenomena*'s popularity partly explains the increasing output of astronomical and astrological treatises in the Hellenistic era, including those by Dorotheus of Sidon (25–75 AD), Pitenius (second century AD), Manetho (b. 80 AD), and Vettius Valens (ca. 180 AD). Eratosthenes's *Katasterismoi* (ca. 245 BC) described the constellations and the planets. Petosiris (ca. 140 BC) connected astrological events to the art of divination, namely what is portended from eclipses, comets and other celestial spectacles. Imbrasius (ca. 25 BC–25 AD) forecasted medical prognoses from stars and weather signs. Similarly, Geminus's *Phaenomena* (ca. 50 AD) analyzed signs derived from weather.

sub pace vacat tantum).[47] Both projects require time for research, study, and thought. Manilius, however, takes pleasure in his *Astronomica* (1.13: *iuvat*). One hopes that Mela also enjoyed composing *De Chorographia*, but he forefronts the labor, effort, and worthiness of his topic (*dignissimum*). For Manilius, his topic is also worthy (1.41: *dignata*). Just as Manilius lauds Augustus, the first citizen and father of his country, whom all the world obeys, and who inspires the poet, Mela glorifies Claudius, "the greatest of emperors" (3.49: *principum maximus*). Imperial encomium was, of course, a common literary trope, perhaps here nothing more than the flattery and gratitude of appreciative clients towards their generous patrons. But we wonder, was Mela writing in response to Manilius? Did he aim to produce a terrestrial treatise that would complement Manilius's *Astronomica*? Although clear verbal echoes are limited, the unusual structural principle of Manilius's geographical ecphrasis suggests that Mela was aware of Manilius and that he chose to model the organization of his world on a Latin rather than Greek archetype. Under Mela, geography becomes a topic for the Roman pen.

Authorial Voice

Mela also views himself, in part, as culturally Roman, and his authorial voice is evident in explicit self-referential comments, such as the use of first and second person verb endings, personal pronouns, temporal remarks, and criticism of sources.[48] Such statements help the reader determine the extent to which an author may have participated in events and how authorial experience and point of view color the presentation of geographical data, history, and contemporary politics. We aim to determine what piqued Mela's curiosity, an author who never mentions himself by name but only obliquely alludes to himself.

Iberia and the Pillars of Hercules provide the geographical nucleus to which Mela is congenitally connected. In his overview of the inhabited world, as our author works around the Mediterranean, he merely mentions that Italy is situated between the Adriatic and the Tuscan Seas, but his account of the coastlines of Gaul and Iberia is pleonastic. Iberia stretches westward and northward, and

47. The glorification of a peace which had been established after prolonged civil war was a common theme in Augustan literature and art. See Paul Zanker, *The Power of Images in the Age of Augustus*, trans. Alan Shapiro (Ann Arbor: University of Michigan Press, 1988).

48. Katherine Clarke, *Between Geography and History: Hellenistic Constructions of the Roman World* (Oxford: Clarendon, 1999), 33–36, 193–244, has examined the evidence for authorial self-representation in Strabo, and Gerhart Grüniger, "Untersuchungen zur Persönlichkeit des älteren Plinius: Die Bedeutung wissenschaftlicher Arbeit in seinem Denken" (PhD diss., University of Freiburg, 1976) has treated authorial persona and self-representation in Pliny the Elder's *Natural History*.

Gaul extends up reaching from "our shores."[49] It is likely here that by "our shores" (*nostris litoribus*), Mela's perspective is Tingenteran. Further, at the end of the third book, the waterway bounded by the Pillars is "our strait" (*nostrum fretum*). Again the perspective could be Tingenteran rather than Roman. Mela lays claim to the Pillars of Hercules because it is, literally, in his backyard. For the Romans, the Pillars of Hercules marked the end of the Mediterranean's well-known and relatively safe waters, and, as the gateway into the vast expanse of unexplored Ocean, it resonated as a powerful symbol of the unknown.[50]

Mela selectively includes places of interest, whether to himself or to his readership, we cannot be certain. From the Roman point of view, some places are inhabitable or not, and some locales are particularly well-known or famous (*clarissimus, notissimus*) or especially wealthy (*opulentissimus*), as emphasized by the use of superlative adjectives. The chorographer employs some form of "greatest/especially" (either as an adjective or adverb: *maximus/maxime*) fifty-one times in his short, hyperbolic treatise. Mela omits lesser known sites and features, as well as locales of historical importance and imperial interest, passing entirely over prime battle sites in Germany, Dacia, the Danube, and the Alpine regions.[51] Actium is merely a name in a list (2.54). Such sites may have held interest for an earlier (Augustan) readership, but Mela has his own agenda. Nonetheless, he does include subtle nods to Claudius's Julian ancestors. For example, he mentions the sanctuary of Venus founded "especially" (*maxime*) by Aeneas on Mt. Eryx in Sicily (2.119). Antandrus in Aeolis is so called either because it was settled by refugees from Andros or because Aeneas's son Ascanius, seized by Pelasgians, had ransomed himself in exchange for the polis

---

49. Mela, *Chor.* 1.19: *Haec in occidentem diuque etiam ad septentrionem diversis frontibus vergit. Deinde rursus Gallia est longe et a nostris litoribus hucusque permissa.*

50. To the contrary, the waters beyond the Pillars of Hercules had been explored by the Phoenicians perhaps as early as the tenth century BC when they were settling that area. The Pharaoh Necho II sponsored a circumnavigation of Africa from the Red Sea up to the Strait of Gibraltar: Herodotus, *Hist.* 2.10; Yaacov Kahanov, "Ma'agan-Michael ship (Israel)" in *Construction navale maritime et fluviale: Approches archéologique, historique et ethnologique*, ed. Patrice Pomey and Eric Rieth; *Archaeonautica* 14 (Paris: CNRS, 1999), 155–60; Duane W. Roller, *Through the Pillars of Herakles: Greco-Roman Exploration of the Atlantic* (London: Routledge, 2006), 23–26. Pytheas of Massilia (flourished 340–290 BC) ventured into the north Atlantic: Strabo, *Geogr.* 2.4.1–2; Roller, *Through the Pillars*, 57–91. The Romans, however, did not trust the waters beyond the Pillars, and Claudius's men nearly mutinied before embarking to cross the English Channel in AD 43 because, so they thought, they were being sent to campaign outside the known world (Cassius Dio, *Roman History* 60.19). For Phoenician settlement of the western Mediterranean, see Ana Arruda Margarida, "Phoenician Colonization on the Atlantic Coast of the Iberian Peninsula," in *Colonial Encounters in Ancient Iberia*, ed. Michael Dietler and Carolina López-Ruiz (Chicago: University of Chicago Press, 2009), 113–30; Duane W. Roller, "Phoenician Exploration," in *Oxford Handbook of the Phoenician and Punic Mediterranean*, ed. Carolina López-Ruiz and Brian Doak (Oxford: Oxford University Press, 2019), 645–53.

51. William H. Stahl, *Roman Science: Origins, Development, and Influence to the Later Middle Ages* (Madison: University of Wisconsin Press, 1962), 69.

(1.92).[52] Mela also mentions the Trojan ally Rhesus, king of Thrace (2.24),[53] and the tomb of the Trojan queen Hecuba on Cynos Sema (2.26).[54] Thus the reader is reminded of Rome's Trojan heritage and Augustus's cultural program that had elevated Aeneas from an obscure Trojan hero to the founder of the Roman race.[55] Mela is also our only source for a "Tower of Augustus memorable because of its inscription" on Iberia's Atlantic coast (3.11).[56] Many imperial monuments were sprinkled throughout the imperial landscape, but our author cites only one, an Augustan memorial in Mela's native territory, reminding the reader once again of his regional perspective.

Mela's *Chorography*, furthermore, reflects the intellectual climate in which he lived. Like Agrippa before him and Pliny afterwards, Mela was an equestrian writing a semi-technical manual at a time when such works flourished. These treatises, moreover, were composed primarily by men of the equestrian rank, to whom largely fell the practical details of provincial administration and who may have lacked the anti-banausic prejudice ingrained in the elite class.[57] In the technical writings of equestrian authors, theory is usually suppressed. Strabo, whose work is suffused with Stoic doctrine,[58] adhered to a multidisciplinary approach to geography. To Strabo, a complete understanding of geography comes only with knowledge of all branches of human knowledge: astronomy, terrestrial history, topography, biology, and philosophy.[59] In contrast, Mela offers no comment on his own philosophy of geography. Like Pliny the Elder, Mela's acquaintance with current philosophical theories was doubtless keen, and awareness of contemporary science helped to shape his outlook in a general

52. See further, Herodotus, *Hist.* 7.42, who examines the origin of the Pelasgians, and Strabo, *Geogr.* 13.1.51, who identifies the inhabitants as Lelagians.

53. An oracle proclaimed that Troy would survive provided that Rhesus's white horses drink the waters of the Xanthus and pasture on the Troad. Diomedes and Odysseus, however, killed Rhesus before the horses were attended: *Il.* 10.432–502; see also Euripides, *Rhesus*; Vergil, *Aen.* 3.14; Ovid, *Metamorphoses* 8.623.

54. See further Strabo, *Geogr.* 8.1.28; Pliny the Elder, *Nat. Hist.* 4.49.

55. Of the Greek players in the Trojan War, Mela mentions Achilles (2.5), Agamemnon's children Iphigeneia and Orestes who came to Tauris where the monstrous xenophobic inhabitants slaughtered newcomers as sacrificial offerings (2.11), Protesilaus whose bones are enshrined at Sestos (2.26: Protesilaus was the first Greek to die at Troy, a datum omitted by Mela), the Cicones (2.28: eschewing mention of Odysseus), and Aulis whence the Greek fleet sailed under Agamemnon (2.45).

56. The tower may have commemorated Augustus's victory over Cantobrio-Asturians in 29–19 BC, possibly at the confluence of the Sars and Ulla Rivers: Adolf Schulten, *Iberische Landeskunde* (Strasbourg: Heitz, 1955), 1:241–42, 357; Silberman, *Mela*, ad loc.

57. Mary Beagon, *Roman Nature: The Thought of Pliny the Elder* (Oxford: Clarendon, 1992), 5.

58. Strabo's tutors included the Stoics Athenodoros from Tarsos, who also taught Octavian (16.4.21), and Boethos of Sidon (16.2.24). Cf. Jérôme Laurent, "Strabon et la philosophie stoïcienne," *ArchPhilos* 71 (2008), 111–27; Alexander Jones, "The Stoics and the Astronomical Sciences," in *The Cambridge Companion to the Stoics*, ed. Brad Inwood (Cambridge: Cambridge University Press, 2003), 342–44; Germaine Aujac, "Strabon et le stoïcisme," *Diotima* 11 (1983), 17–29.

59. Vitruvius had made similar observations on architecture: *On Architecture* 1.1.

way, as for example in his description of the inhabited world or the intellectual culture of India.[60] But, like Pliny, Mela may have taken little interest in philosophical argument for its own sake.

De Chorographia opens with an overview of the work's scope and intent. In the prologue, Mela declares the factual nature of his topic, based on hard data. He treats a dry, tedious, but necessary and important science. With a first person singular deponent verb, he promises to describe the world, listing the names of places and peoples in a rational, orderly fashion (1.1: *orbis situm dicere aggredior*). But this deployment of locales is "intricate" (*perplexo*). Furthermore, documenting the human stamp upon the earth is a demanding topic (*impeditum opus*). Nonetheless, pursuing and understanding the geography of the earth is "well worth" the effort (*dignissimum*). The data that Mela intends to include are "superlatively clear" (*clarissima*) or well-known or interesting. In his restatement of method, the superlative "most suitable" (*commodissima*) highlights the advantage of starting at the beginning, not with Hispania in contrast with Strabo and others, but with Mauretania (1.25; 2.97). Although Mela does not explain the counter-clockwise exegesis around the Mediterranean, the superlative adjective emphasizes his point of view. Upon first glance, the agenda is straightforward and factual. The book, as the author himself warns, will not be a page-turner. The genre makes specific stylistic demands, such as "repetitiousness and the necessary inclusion of catalogues."[61] Geography is a topic that is superlatively (*minime*) unsuited to eloquence. Employing a common rhetorical trope, Mela avers his own rhetorical talent but in a very impersonal way (1.1: *si non ope ingenii orantis*, "if not by the power of the speaker's talent"). Nonetheless, by denying his own talent Mela broadcasts his literary aspirations. The use of the passive-looking deponent verb, with its middle force (*aggredior*: "I approach for myself"), may serve to distance Mela from both material and readership in order to establish the appearance of objectivity. Throughout the text, Mela utilizes the passive voice liberally, even to describe strictly physical phenomena: on the Mauretanian side of the Pillars of Hercules, the sea *is poured* out (by whom or what?) rather broadly (1.27: *hic iam mare latius funditur*); the southern part of Italy *is cut* into two horns (2.58: *finditur*). The use of the passive voice emphasizes that the knowledge is common, requiring no footnote. Mela further implies objectivity in his prologue with the impersonal "it is agreed" (*constat*), further distancing himself from text and reader in employing two passive infinitives to describe the value of his program, which is "very worthy *to be seen and to be understood*" (1.1: *verum aspici tamen cognoscique dignissimum*).

---

60. Beagon, *Roman Nature*, 15.
61. Romer, *Mela*, 9.

But Mela is also an actively engaged writer who takes responsibility for his work. With the use of first person singular verbs (instead of the more formal and common first person plural verbs), he immediately establishes himself, in a very personal way, as the author: "*I* undertake" (*aggredior*); "*I* will say" (*dicam*); "*I* will explain" (*expediam*) (1.1–2).

Mela also uses a verb in the first person plural to reiterate the geographical agenda of *De Chorographia*. Near the end of the second book, the geographer returns to his starting point, the Pillars of Heracles, which he now describes more fully (2.95: *ut initio diximus*; see also 1.272; cf. Strabo, *Geogr.* 3.5.5). Once Mela has completed his circuit of the Mediterranean, returning to the Pillars of Heracles, he is careful to remind his reader of the sequence laid out in his prologue, just as the island Gades, charmingly personified, had reminded him to do (2.97: *Gades admonet*).[62] This is the slice of the inhabited world that Mela knows best, and his account of it is detailed, intimate, and personal. Our guide will then describe the islands before proceeding to the lands that border the Atlantic, as he had promised at the beginning (2.97: *ut initio promisimus*). Finally, Mela employs a first person plural verb to conclude both the journey and the treatise, returning to the Pillars of Hercules, the point where "we made our start," in an elegant ring structure (3.107: *unde initium fecimus ... in Nostrum iam fretum vergens promontorium*).[63] The Pillars, significantly near Mela's birthplace, mark the beginning, midpoint, and end (*terminus*) of both the journey and the text, providing the armature of *De Chorographia*.

In narrative literature, the second person singular is rare, applied for striking and often poignant affect.[64] Twice in the second book and once in the third, Mela addresses his reader directly (2.78: *credas*; 2.89: *legas*; 3.40: *intuearis*). The reader is invited to become not just a passive bystander but a fellow traveler on the journey through the Mediterranean world: you, dear reader, might pick out your way along the coasts of Iberia (2.89: *legas*). Mela does not mark out the cities along his coast as mere points on a map but as potential destinations for a real visitor, the reader, who has joined the author on a literary trip back to his very homeland. Whereas Romer argues that Mela invites individual assessment,[65] we suggest that Mela implies his own willing suspension of disbelief as well as that of his readership with a jussive subjunctive: you, gentle reader, "would believe" (or "could believe") (*credas*) the story about Hercules fighting the sons of Neptune on the Rocky Beach on the Rhone [*Litus Lapideum*; the Crau], a site that is remarkable for its numerous, widely scattered stones. When Hercules

---

62. See Romer, *Mela*, 10.

63. Playfully, "end" (*terminus*) is the final word of the treatise.

64. As, for example, when Homer addresses Odysseus's swineherd Eumaios in the second person: "O swineherd, you said ..." (προσέφης: *Od.* 14.55).

65. Romer, *Mela*, 14.

ran out of arrows, Jupiter sent a rainstorm of rocks to help his son. The rocks remain at the site as incontrovertible, physical proof that lends credence to the myth. Thus, the story is plausible, and Mela invites the reader to accept its veracity. Finally, with a second-person deponent singular verb, "you might see for yourself" (*intuearis*), Mela suggests, however keen his reader's powers of observation may be, that the course of the river Araxes in the Taurus mountains eludes detection because its flow is so "calm and quiet" (*placidus et silens*). Although you, gentle reader, might watch it very closely, so Mela intimates to his reader, you will be unable to observe the current of the placid Araxes River as it debouches into the sea after its long and violent journey from its source.

Mela frequently separates himself from text and reader with impersonal formulae or expressions of skepticism in the third person. When listing the bizarre, barely human creatures who inhabit the interior of the African continent—Aigipanes (Goat-Pans), the Blemyes whose faces are on their chests, the nude, pacifistic Gamphasantes, and the Satyrs—Mela expresses disbelief with an impersonal expression: "if it is agreeable to believe" (1.23: *si credere libet*).[66] Mela describes this quartet more fulsomely later in the same book but without innuendo (1.47–48). In the third book, Satyrs and Goat-Pans are once again cited, identified as the owners of the extensive fields beyond the fiery Chariot of the Gods (3.94).[67] Satyrs and Goat-Pans are mentioned thrice in three distinct sections, close to the beginning and the end of the work, an indication of Mela's fascination with these exotic hybrids. The Goat-Pan is obliquely evocative of the Capricorn, which was advanced by Augustus as his own sign and extolled by Manilius as protecting the *princeps*.[68] Octavian employed the Capricorn as early as the late 40s BC in order to emphasize his role as the avenger of Caesar's murder and later as the savior of the state against Marcus Antonius.

The constellation Capricorn was also lauded by Publius Nigidius Figulus, Cicero's Pythagorean friend, as a symbol of enlightened political rule and civic order (Nigidius Figulus, *Sphaera Graecanica* 122–25).[69] According to Nigidius,

---

66. For Goat-Pans, see further Pliny the Elder, *Nat. Hist.* 5.7, 48. Strabo situates the Blemyes along the Nile: *Geogr.* 17.1.2. The Gamphasantes seem to be Herodotus's Garamantes (4.183–184; see also Pliny the Elder, *Nat. Hist.* 5.45). Diodorus locates the Satyrs in Ethiopia (*Hist.* 1.18), and Pliny situates them in India (*Nat. Hist.* 7.24). See also Pliny the Elder, *Nat. Hist.* 6.197; Silberman, *Mela*, ad loc.

67. Theon Ochema, identified with Hanno's volcanic mountain, is possibly Mt. Kakulima in Guinea: Silberman, *Mela*, ad loc.

68. Manilius, *Astron.* 2.508–509: *contra Capricornus in ipsum convertit visus. Quid enim mirabitur ille maius, in Augusti felix cum fulserit ortum?*

69. Tonio Hölscher, "Ein römischer Stirnziegel mit Victoria und Capricorn," *JRGZ* 12 (1965), 59–73;; Eugene Dwyer, "Augustus and the Capricorn," *MDAI(R)* 80 (1973), 59–67; K. Kraft, "Zum Capricorn auf den Münzen des Augustus," *JNG* 17 (1967), 17–27. On the day of Octavian's birth, Nigidius had presumably foretold that the child would rule the world (Suetonius, *Augustus* 94).

Capricorn, the goat-fish, represents Pan, who had suggested that the Olympians assume animal disguises in order to deceive Typhon. Before hiding in the Nile, Pan disguised himself as a composite creature that synthesized the characteristics of both god (Pan the goat) and his means of evasion (fish in the Nile) (cf. Manilius, *Astron.* 2.167–72). Pan's transformation into the Capricorn was key in preserving Olympian rule and thwarting Typhon's tyranny. Likewise, Octavian, who was protected by this sign, restored political and military concord to a stable, united Rome, and he utilized the Capricorn as a symbol to emphasize security and peace on both land and sea.[70] Like Pan, who had safeguarded the Olympians in the medium of water (the Nile), Octavian had saved the Roman state in a decisive naval battle at Actium. Octavian's sign, the Capricorn, which held sway over the sea, came to represent that maritime victory which was emphasized in coin releases and public art.[71] Nowhere does Mela mention the Capricorn, but three times he reminds his readers of Pan, a composite god associated with the Capricorn, a symbol that becomes particularly Roman and nearly synonymous with the first *princeps* and an end to civil war at Rome.

Mela's Interests

Stahl sternly criticized Mela for including *paradoxa* while omitting politically significant sites, but *De Chorographia* is a short work (three books, just under 16,500 words) with an overly ambitions agenda (a description of the inhabited world as known to the Romans in the first century AD).[72] It is necessarily selective, and what Mela includes may be more illuminating and interesting than what he has omitted. Mela describes many places as "famous," "well known," or "worthy" without explanation. For example, Atthis (Attica) is the most famous place in Greece, cited, however, not by its common toponym but rather by the name of an obscure mythological character, Atthis, a daughter of Kranaos, who gave her name to the region (2.39).[73] Our author has no interest in repeating old information; he wants to offer something fresh, which he expresses in an innovative way.

70. The Capricorn was widely used on Augustan coinage, public monuments, and as a *signum* for military units: e.g., *Legio II Augusta*, whose *dies natalis* fell on Augustus's: *CIL* VII 103, *addit.* 306 (*RIB* 327); cf., Lawrence Keppie, *The Making of the Roman Army: From Republic to Empire* (Norman: University of Oklahoma Press, 1998), 139–40; Graham Webster, *The Roman Imperial Army of the First and Second Centuries A.D.*, 3rd edition (Norman: University of Oklahoma Press, 1998), 136. See also Simonetta Terio, *Der Steinbock als Herrschaftszeichen des Augustus* (Münster: Aschendorff, 2006), 97–110.

71. Zanker, *Power of Images*, 82–85; see also Germanicus, *Aratea* 554–560, for the Capricorn and Augustus's apotheosis.

72. Stahl, *Roman Science*, 69.

73. Cf. Pseudo-Apollodorus, *Bibliotheca* 3.14.5; Strabo, *Geogr.* 9.1.18; Pausanias, *Descr.* 1.2.5.

We consider where Mela pauses to add a detail, however scant, regarding the places he lists, including famous inhabitants, important events in history or myth, natural wonders, and *paradoxa*. Mela is impressed by the very large and the very old. He singles out lofty mountains, including Mt. Casius in Arabia, on whose summit sunrise is visible from the fourth watch onward (1.61), as well as Mt. Athos (in the northern Aegean), whose peak rises above the point where rain falls (proved by the accrual of sacrificial ashes which ordinarily should wash away with the rains: 2.31).[74] As proof of the sea-monster slain by Perseus, Mela offers the huge bones found at Iope on the Judean coast (1.64). He also records the length of India's coastline as a course of "sixty days and nights for sailors" (3.61: *per sexaginta dies noctesque*).[75]

Mela also singles out Xerxes's unusual, bold bridge over the Hellespont as a marvelous and enormous (*ingens*) deed, but he omits the Persian king's name (2.26: *mirum atque ingens facinus*; see also Herodotus, *Hist.* 7.59–60). Xerxes is elsewhere mentioned by name twice (2.28, 32).[76] Mela also makes note of Cambyses (1.64), Darius III (1.70), and Semiramis whom Mela credits with Babylon's impressive size and the hydraulic infrastructure that enabled the famous rivers there to irrigate the fields (1.63; see also Herodotus, *Hist.* 1.184). Mela distinguishes the Eurymedon River as the site of Cimon's great maritime victory against the Phoenicians and Persians in 468 BC (1.78), Port Coelus where the Athenian fleet was destroyed by the Lacedaemonians in 411 BC (2.26; cf. Thucydides, *Hist.* 8.106), and Salamis where the Persian fleet was vanquished in 480 BC. Mela emphasizes Persian losses instead of the Greek victory, perhaps in order to forefront to his own cultural ancestors, the Phoenicians, who had fought as allies in Xerxes's navy (2.109; cf. Herodotus, *Hist.* 7.89, 96; 8.90). Mela includes Dyrrachium (Illyricum: 2.55), where the sea was pivotal in Pompey's victory over Caesar, but omits Pharsalus (in Thessaly), the site of Caesar's land victory over Pompey, a battle which marked the end of that civil war. Thus, it seems, nautical battles intrigued Mela, whose homeland boasted

---

74. Tourists did climb famous mountains. Guides could be found at Centoripa, a town at the foot of Aetna (Strabo, *Geogr.* 6.2.8), whose famous tourists included Empedocles (Diogenes Laertius, *Lives* 8.69), Seneca the Younger's friend Lucilius (Seneca the Younger, *Epistle* 79.2–3), and the emperor Hadrian (reigned 117–138 AD), who saw there a magnificent sunrise, "many colored, like a rainbow" (*SHA Hadrian* 13.3). During his visit to Syria, Hadrian observed the sunrise from Mt. Cassius where he was nearly struck by lightning, and he climbed a mountain in Pontus (*SHA Hadrian* 13.12). In Pausanias, the tourist finds notes on the trails to the summit of Mt. Parnassus (*Description* 10.5.1; 10.32.7). See also J. Donald Hughes, *Environmental Problems of the Greeks and Romans: Ecology in the Ancient Mediterranean*, 2nd edition (Baltimore: Johns Hopkins University Press, 1994), 55–56.

75. Whenever possible, sailing was restricted to daylight hours.

76. At Doriscos Xerxes estimated his troop strength by space because there were too many to count (cf. Herodotus, *Hist.* 7.59–60).

a long maritime heritage.[77] Whereas Rome was more famous for its army than its navy, Mela spotlights naval expeditions, perhaps in a subtle effort to exalt Claudius whose "conquest" of Britain rendered Ocean, the Atlantic, as finally conquered (Suetonius, *Claudius* 17: *quasi domiti Oceani insigne*).[78] Also meriting mention by name are Alexander (1.66, 70, 98; 2.34), Juba II (1.30), Cato the Younger (1.34) who died at Utica (a Phoenician settlement), and the Philaeni brothers (1.38) who had been sent from Carthage (also a Phoenician settlement) to broker peace with the Cyrenaeans. Finally, Mela mentions Pompey who gave Pompeiopolis/Soloe in Cilicia to the pirates in 66 BC (1.71), and who had campaigned in Hispania against Sertorius in the 70s BC.[79] Mela is also eager to point out his own birthplace (Tingentera), as well as those of men he admired, including Thales the astronomer, Timotheus the musician, Anaximander the natural philosopher—all from Miletos (1.86)—and Democritus from Abdera (2.29). Mela also reports that Philip and Alexander hailed from Pelle (2.34).

De Chorographia is sprinkled liberally with mythological allusions and references. Aside from the Trojan War and its preliminaries, Mela seems especially interested in the myths of Jason and Hercules. Suppressing the names of more familiar characters, he refers evasively to the myths of Perseus (where the Aithiopian king Cepheus, father of Perseus's bride Andromeda, and the king's brother Phineus are prominent: 1.64), and Bellerophon (elicited only by mention of the Chimaera: 1.80). Jason, whose adventures, naturally, provide fodder and interest to literary geography, is linked with several toponyms: Chalcedon where Jason dedicated a temple to Jupiter (1.101); places along the Phasis River, well known because of Golden Fleece, the object of Jason's heroic quest (1.108); Lemnos, where once only women lived after they had slaughtered their men (Jason's name is withheld as are the sexual exploits of the Argonauts there: 2.106); and Crete, famous for many things including Talus who merits an item in Mela's list (2.112).[80]

---

77. The area had been settled by Phoenicians (see also Livy, *Ab urbe condita* 25.40; Pliny, *Nat. Hist.* 5.24) whose nautical skills were legendary (Herodotus, *Hist.* 3.19; Strabo, *Geogr.* 1.1.6; Pliny, *Nat. Hist.* 7.209).

78. The conquest of Ocean figures into Claudian encomium (*Latin Anthology* 419, 423), and Ocean becomes the new boundary of the Empire (*ILS* 212). Despite never crossing the channel or setting foot in Britain, Gaius declared victory over Ocean (Suetonius, *Gaius* 46: *spolia Oceani*).

79. During the Sertorian War, many Iberian cities had transferred their alliance to Pompey, and the general had also protected towns against Sertorius's sieges (Plutarch, *Sertorius* 18). For his victories in Iberia, Pompey celebrated a triumph in 71 BC, and his triple triumphs (celebrating his victories over Numidia, Iberia, and Asia) were widely acclaimed: Cicero, *For Sestius* 61.129; Manilius, *Astron.* 1.793–794; Valerius Maximus, *Memorable Deeds and Sayings* 5.1.10; Petronius, *Satyricon* 119, 240–241; Pliny the Elder, *Nat. Hist.* 37.13; Plutarch, *Pompey* 40, 45. In Lucan, the triple triumph is Pompey's "greatest joy" (*Pharsalia* 7.685–686; cf. 7.685–686; 8.553, 813–815; 9.177–178, 599–600).

80. Originally sent by Zeus to Crete to protect Europa, the bronze robot hurled rocks at the Argo as it passed his island, and was vanquished by Medea (Apollonius of Rhodes, *Argonautica* 4.1638–1672).

A hero whose adventures also took him to all corners of the Mediterranean world and greatly overlapped with Jason's exploits, Hercules had strong associations with the western Mediterranean in general and Iberia in particular. In Hispania Ulterior, where he is widely attested on coins, Hercules was assimilated with the Phoenician Melkart, the divine ancestor of the Tyrian royal family. Near Gades was the magnificent temple of the oracular and popular Hercules-Gaditanus. This is presumably the temple of the "Egyptian Hercules" in Mela, founded by Tyrians and where Hercules's bones are buried (3.46; see Strabo, *Geogr.* 3.5.5–6).[81] Our Iberian author is careful to distinguish between Greek and western incarnations of the demigod. As a "local hero," Hercules's resonance with Mela was vigorous, and Mela repeatedly reminds his readers of the hero's deeds, overtly and circuitously citing his legend at least thirteen times. Tinge, the very first town mentioned by Mela, was founded long ago by the giant Antaeus (a wrestler-son of Poseidon whom Hercules grappled to death),[82] and Antaeus's tomb, which resembles a small hill, is the last monument that the reader will visit (3.106). Thus Mela layers a mythic ring-structure onto his Ibero-centric geographical framework. As a point of local pride and geological history, Mela also recounts how Hercules created his eponymous strait (1.27). Along the way, Mela shows his reader the city of the Mariandyni, a *polis* on the Pontus founded by the Argive Hercules (the fact that it was later called "Heraklea" lends credibility, so claims the author) and the Acherousan cave, where Cerberus was dragged up from the underworld (1.103; Hercules's name need not be cited). In northern Thrace, Hercules fed the xenophobic Diomedes to his vicious man-eating horses (2.29), and Lusitania was the home of Geryon, the composite monster whose red cattle Hercules had famously stolen (3.47).[83] Mela alludes to a liaison between Hercules and Echidna (2.11),[84] and the obscure fight between Hercules and Neptune's sons, Alebion and Dercynos, at Litus Lapideum (2.78). King Erythras, buried at Ogyris ("more famous than" other Arabian islands), may have been a son of Hercules (3.79).[85] Finally, the reader visits the sites where the Greek Hercules died, Mt. Oeta (2.36: *Grai Herculi*), and where his bones are interred (one of the promontories of the eponymous strait: 3.46).

---

81. Simon J. Keay, *Roman Spain* (Berkeley: University of California Press, 1988), 147.

82. For proof of Antaeus's residency Mela offers an elephant-hide shield that is too large for contemporary residents to use easily.

83. See also Apollodorus, *Library* 2.5.10.

84. According to Herodotus (*Hist.* 4.8–10), Echidna had ransomed Herakles's horses in exchange for sex. Their progeny included Skythes, the eponymous founder of the Scythian race.

85. See also *Chor.* 3.72 where an Erythras ruled near the Erythraean [Arabian] sea, thus explaining the toponym. For Erythras as a son of Herakles: Apollodorus, *Library* 2.7.8. Poseidon also had a son Erythras (Scholia on Homer, *Il.* 2.499), and Pausanias (6.21.11) listed an Erythras among Hippodamia's suitors. Pliny follows Mela (6.153), but Strabo, *Geogr.* 16.3.5, attributes the monument to Alexander's admiral Nearchus.

Mela is also fascinated by *paradoxa* of the natural world, including strange and often dangerous beasts whose presence can render even a fertile territory unlivable by humans. For example, despite the fertility of the soil, savage griffins make the lands beyond the Rhipaean mountains uninhabitable (2.1). Mela also observes that coastal Gaul, a rich, healthful (*salubris*) land is "rarely crowded with harmful creatures" (3.17: *noxio genere animalium minime frequens*). Thus, the absence of harmful animals contributes to making a place healthy. The territory between the Sacae and Scyths is uninhabitable because of wild animals (3.59), and the parched middle zone of Africa's western coast is filled with sand and snakes (3.100).[86] Mela thus sees a clear division between cultivated and feral territory: savage griffins inhabit the trans-Rhipaean regions, and serpents possess the torrid middle zone in Africa. And it is dangerous for men to encroach into the territory of wild animals. In the Scythian forests, for example, swift Hercynean tigers will stealthily track hunters who have stolen their cubs, retrieving them, one at a time, until the thief reaches the safety of a densely populated area (3.43).

Mela offers subtle grammatical and syntactical clues of his interests. Although he simply lists many sites, monuments, and *paradoxa*, others he discusses at length, including Psammeticus's labyrinth on the floating island of Chemnis (1.55–56) and Typhon's cave (1.76). Mela generously employs the future passive periphrastic construction to indicate something that "ought to be" recounted (e.g., *memorandus, dicendus, referendus*). These gerundives may merely reflect Mela's attempts at stylistic *varatio*, or they may cue what especially sparked the author's curiosity. Among the items that "must be" described are the well known: the Mausoleum of Halicarnassus, one of the seven wonders of the ancient world, completed by Artemisia in 351 BC (1.85);[87] and the phoenix, the unique bird that spontaneously self-generates neither through sex or birth (3.83), recalling the parthenogenetic gorillae of Hanno's expedition.[88] Worthy of report, in Ethiopia, is the tiny *catoblepas* (who casts its gaze downwards). This unusual creature who can hardly support its heavy head, despite its gentle manner, causes death to all those who gaze into its eyes (3.98), a natural analog to the legendary Gorgons whose islands are nearby.[89] Intriguing is the funerary

---

86. We recall that Manilius also characterizes Libya as sheltering "horrible serpents," "gigantic elephants," "ferocious lions," and "hideous monkeys," and he remarked also on the dry sands and the land's hostility (*natura infestat*): *Astron.* 4.662–669. Mela's *infesta seperpentibus* echoes Manilius's *natura infesta.*

87. See also Strabo, *Geogr.* 14.2.16; Pliny the Elder, *Nat. Hist.* 36.30; Pausanias, *Descr.* 8.16.4. To Mela, the structure seems to be interesting only by virtue of its position among the seven wonders.

88. For the phoenix, see also Herodotus, *Hist.* 2.73; Ovid, *Metamorphoses* 15.392–407; Pliny the Elder, *Nat. Hist.* 10.3–5; Martial, *Epigrams* 10.16; Tacitus, *Annals* 6.28.

89. Athenaeus, *Deipnosophists* 5.221b describes the animal as a "gorgon" whose eyes are covered by long hanging hair and whose breath can also kill. Fighting against Jugurtha in 117 BC,

monument of the poet Aratus, where rocks hurled at the monument happen to shatter (1.71). Mela explains neither why people throw rocks nor how they come to shatter. Also worthy of remembering is the light-devouring cave of Typhon, never easy to investigate owing to its very narrow passages and because it kills all things sent into it, as reported by those who have experienced it, on Mela's report (1.76).[90]

Finally, Mela occasionally pauses for lengthy, vivid descriptions of *paradoxa* and natural wonders. He painstakingly relates the unusual behavior of the fountain of the sun which boils at midnight but grows colder as the sun rises, icing over at high noon (1.39). Meriting extensive description are the paradoxical sunrises at Mt. Ida which appear first as scattered flames until the lights eventually coalesce to form the sun as seen elsewhere (1.94). Mela also provides a meticulous account of a beautiful Corycian cave, which grows more impressive the deeper one descends. Eventually coming upon an embedded cave, the visitor might be startled by a "divinely-sent sound of cymbals and the great din of rattling." The spelunker finally reaches a mighty underground river (1.72–75). Mela provides unusually rich data on the navigable, calm Borysthenes, the loveliest of the Scythian rivers, whose delicious waters nurture enormous, flavorful, boneless fish (2.6). Mela also fastidiously recounts the violent behavior of the Araxes River, as it forces its way through narrow cliffs, growing choppier and so swift that it does not cascade down where the terrain dips but instead "projects its wave out (horizontally) beyond its channel" (3.40).

Mela's attention to *paradoxa* reflects the aesthetic tastes of his class and epoch, when grotesque hybridism was in vogue, as evidenced by the popularity of Third Style Wall Painting (characterized by otherworldly, delicate architectural details and fantastical animals and landscapes) and the regard for dwarves and hunchbacks as "pets."[91] We recall Mela's particular interest in the hybrid Goat-Pans and Satyrs. Pliny the Elder may summarize the first-century AD attitude to so-called *paradoxa* with his assertion that "the more I observe nature, the less prone I am to consider any statement about her to be impossible" (*Nat. Hist.* 11.6). As geographical knowledge accrued and increasingly strange reports of bizarre peoples, plants, and animals came to Rome, "the strangeness of what had been discovered heightened expectations of what was to come."[92]

Marius's troops presumably encountered such a creature, but they were killed by its gaze when the startled creature shook the hair away from its face and looked at soldiers.

90. See also Pindar, *Pythian Ode* 1.32; Strabo, *Geogr.* 13.4.6.

91. Ian M. Barton, *Roman Domestic Buildings* (Exeter: University of Exeter Press, 1996), 85.

92. Beagon, *Roman Nature*, 10. The Greeks had long believed that Africa was a source of wonders: Aristotle, *History of Animals* 606b20 ("always something fresh in Libya"). See also Pliny the Elder, *Nat. Hist.* 8.42: "Africa always brings something new" (*semper aliquid novi African adferre*).

Mela from Hispania Baetica

Mela withholds personal information, with the exception of the name of his hometown and the era in which he lives. He hails from Tingentera, a city settled by Phoenicians, perhaps from Tinge (Tangiers, hence *Tinge*ntera), on the coast of Hispania Baetica near the Strait of Gibraltar (2.96: *unde nos sumus, Tingentera*). His regional pride is evident. Mela describes Iberia exuberantly, in comparison with his almost dismissive treatment of Italy: "since the order of the work demands it rather than because it needs to be pointed out, a few things will be mentioned about Italy" (2.58: *De Italia magis quia ordo exigit quam quia monstrari eget, pauca dicentur*). In comparison with the effusive accounts of Italy by Strabo (*Geogr.* 5.3.5) and Pliny (*Nat. Hist.* 3.40), hyperbole is starkly lacking in Mela's Italian treatment. Mela applies one superlative adjective to Italy in his list of the wealthiest inland cities (2.60: *opulentissimae*), and he concedes that Campania is "pleasant" (2.70: *amoena*).

The hyperbole with which Mela describes his home province in general, and his hometown (Tingentera) in particular, anticipates Pliny the Elder's ebullient treatment of Italy and Rome (*Nat. Hist.* 3.40). Whereas Strabo aims to flatter a Roman readership and Pliny intends his collection of "20,000 useful facts" (*Nat. Hist.* preface 17) to glorify Rome, Mela may have regarded the city as so well known by his target readership as to require no description at all (2.58: *nota sunt omnia*). He preferred, perhaps, to avoid banality by not burdening his reader with the familiar,[93] or he may have intended to pen a more detailed account of the city in future (2.60).[94] In contrast, Mela employs five superlative adjectives to describe Iberia, where even topographical features are unparalleled, horses and precious metals abound, and the land is so fertile that flax and esparto grow even during droughts (2.86). Rome is little more than an item in a list of wealthy Italian cities: with a single passive participle, we learn that the city was founded long ago by shepherds (2.60: *et Roma quondam <a> pastoribus condita*).

Mela the Roman Author

In spite of his palpable regional pride, Mela aims to synthesize his Iberian identity with his position as a Latin author in Claudius's court at Rome. Our

---

93. Perhaps just a *topos* of the genre (Pliny the Elder, *Nat. Hist.* 3.40: *sed quid agam?*), Mela's *pauca dicentur* recalls the tone that he employed for Athens (2.41: *clariores quam ut indicari egeant*), Italy, and New Carthage (2.94: *nihil referendum est*).

94. *Nunc si pro materia dicatur alterum opus* ("if another work should be designated for the material"); cf. Romer, *Mela*, 24n37, 86n55. Mela's use of a subjunctive verb (*dicatur*) in a future less vivid conditional renders his proposed book on Rome as merely a remote possibility.

writer uses language, grammar (first person verbs), and first person personal pronouns (we/us) to distinguish between "us" (who are culturally Roman or Graeco-Roman) and "them" (who are culturally non-Roman). *We* inhabit this part of the world (1.4: *incolimus*), whereas the Antichthones inhabit another part of the world. The Indians use elephants as easily as *we* (Greeks and Western Mediterranean peoples) use horses (3.63: *ut elephantis etiam et ibi maximis sicut nos equis facile atque habiliter utantur*).

Mela further emphasizes this otherness in his use of toponyms and ethnyms. He clearly recognizes that a place may be known by several names, according to how many cultures inhabit, use, or know of a place. For example, seven separate ethnyms designate those who dwell around the Caucasus mountains: Cerauni, Taurici, Moschi, Amazonici, Caspii, Coraxici, and Caucasii. Each ethnic group, moreover, has its own name for the mountains (1.109: *ut aliis aliisve adpositi gentibus ita aliis aliisque dicti nominibus*). Mela occasionally gives Greek names for well-known places: the sea that *we* call Tuscan is the one which the Greeks call Tyrrhenian (1.17: *nos Tuscum [quem] Grai Tyrrhenicum perhibent*). Mela generally withholds non Greco-Roman place-names but nonetheless brings attention to the fact that the Roman name for a location often differs from other names: we (Romans) call the landmass on one side of the Mediterranean *Africa*, as distinct from the Greek toponym *Libya*, omitted here by Mela (1.8: *vocamus*). The Greeks call one of their Thracian settlements *Macron Teichos*; Mela omits the Roman name, if there was one (2.24). Again, we call a certain Hibernian promontory "Celtic Point" (3.9). In Mela's text it is unclear if the perspective is Roman or Celtic. Although the ethnym is first attested in Greek,[95] by Caesar's day inhabitants of some parts of Gaul were self-identifying as "Celtic" (Caesar, *Gallic War* 1.1: *ipsorum lingua Celtae*). Mela omits other (local?) names. Furthermore, some local toponyms are altogether unpronounceable in the Latin language, such as the native names for the rivers in the Cantabri territory in northern coastal Iberia (3.15: *quorum nomina nostro ore concipi nequeant*).[96]

Mela the Roman Stylist

Mela's *oikoumene* aligns substantially with the old Greek model where the known, habitable world was a zone stretching from east to west, entirely

---

95. "Keltoi" is first attested in Herodotus (*Hist.* 2.33.3). See also Xenophon, *Hellenika* 7.1.20; Strabo, *Geogr.* 4.1.1. "Celt" does not survive as an ethnym in Old Irish.

96. Pliny the Elder claimed to use non-Latin words only through compulsion (*Nat. Hist.* praef. 13: *aut rusticis vocabulis aut externis, immo barbaris etiam, cum honoris praefatione ponendis*).

surrounded by Ocean, organized into three major landmasses—Asia, Europe, and Libya (Africa)—separated by bodies of water, the Nile and Phasis Rivers and the Mediterranean and Black Seas. This paradigm, of an *oikoumene* divided into symmetrical landmasses, was formalized by Anaximander and Hekataios (ca. 550–500 BC); in Herodotus we find the first extant documentation of a tripartite *oikoumene* (*Hist.* 4.42).[97] Despite the development of empirical cartography and the accrual of geographic data gathered in the wake of Alexander's campaigns, the tripartite model endured, and this design found expression in Latin geographical writers. The tripartite model is applied even to portions of the world, such as to Julius Caesar's Gaul (*Gallic War* 1.1.1: *Gallia est omnis divisa in partes tres*) and Mela's Hispania (2.87).

Mela's description of Hispania strongly recalls Caesar's geography of Gallia (2.87).[98]

Julius Caesar, *Gallic War* 1.1

Gallia est omnis divisa in partes tres, quarum unam incolunt Belgae, aliam Aquitani, tertiam qui ipsorum lingua Celtae, nostra Galli appellantur. Hi omnes lingua, institutis, legibus inter se differunt. Gallos ab Aquitanis Garumna flumen, a Belgis Matrona et Sequana dividit. Horum omnium fortissimi sunt Belgae, propterea quod a cultu atque humanitate provinciae longissime absunt, minimeque ad eos mercatores saepe commeant atque ea quae ad effeminandos animos pertinent important, proximique sunt Germanis, qui trans Rhenum incolunt, quibuscum continenter bellum gerunt. Qua de causa Helvetii quoque reliquos Gallos virtute praecedunt, quod fere cotidianis proeliis cum Germanis contendunt, cum aut suis finibus eos prohibent aut ipsi in

Pomponius Mela, *Chor.* 2.87–88

Tribus autem est distincta nominibus, parsque eius Tarraconensis, pars Baetica, pars Lusitania vocatur. Tarraconensis, altero capite Gallias altero Baeticam Lusitaniamque contingens, mari latera obicit Nostro qua meridiem, qua septentrionem spectat Oceano. Illas fluvius Anas separat, et ideo Baetica maria utraque prospicit, ad occidentem Atlanticum, ad meridiem Nostrum. Lusitania Oceano tantummodo obiecta est, sed latere ad septentriones, fronte ad occasum. Vrbium de mediterraneis in Tarraconensi clarissimae fuerunt Palantia et Numantia, nunc est Caesaraugusta; in Lusitania Emerita, in Baetica Hastigi, Hispal, Corduba.

97. See Clarke, *Between Geography and History*, 113; Duane W. Roller, *Ancient Geography: the Discovery of the World in Classical Greece and Rome* (London: Tauris, 2015), 50–51.

98. The parallels have gone unnoticed by the commentators.

eorum finibus bellum gerunt.
Eorum una pars, quam Gallos
obtinere dictum est, initium capit
a flumine Rhodano; continetur
Garumna flumine, Oceano, finibus
Belgarum; attingit etiam ab Sequanis et Helvetiis flumen Rhenum;
vergit ad septentriones. Belgae ab
extremis Galliae finibus oriuntur,
pertinent ad inferiorem partem flu-
minis Rheni, spectant in septentrionem et orientem solem. Aquitania
a Garumna flumine ad Pyrenaeos
montes et eam partem Oceani quae
est ad Hispaniam pertinet; spectat
inter occasum solis et septentriones.

All Gaul is divided into three parts:
the Belgae inhabit one part, the
Aquitani another; those who are
called Celts in their own language,
but Gauls in ours, occupy the
third part. They all differ amongst
themselves in language, customs,
and laws. The Garumna River
(Garonne) separates the Gauls from
the Aquitani. The Matrona (Marne)
and Sequana (Seine) separate them
from the Belgae. Of these peoples,
the bravest are the Belgae, because
they are most distant from the
splendor and civilization of the
province, and least of all do merchants visit them and import those
things that pertain to enervating the
minds. And they are very close to
the Germani with whom they con-
tinuously wage war. For this reason,
the Helvetii also surpass the other
Gauls in courage because they com-
pete with the Germani in nearly

Moreover Hispania is distinguished
by three names, part of it is called
Tarraconensis, part Baetica, part
is called Lusitania. Touching the
Gauls at one head and Lusitania and
Baetica at the other, Tarraconensis
casts its sides to Our Sea where
it glances southwards, where it
looks north, it borders Ocean. The
Anas River (Guadiana) separates
those regions, and consequently
Baetica gazes forth at two seas:
the Atlantic to the west; Our Sea
to the south. Lusitania is bordered
only by Ocean, with a side to the
north and a front to the west. The
most famous of the inland cities of
Tarraconensis had been Palantia
and Numantia, but now Caesaraugusta (Saragossa) is the most distinguished, in Lusitania, Emerita
(Mérida), in Baetica Hastigi (Ecija),
Hispal (Seville), and Corduba.

daily skirmishes, while either they
expel the Germani from their own
borders or they themselves wage
war in German territory. One part
of these territories, which is said
to extend towards the Gauls, takes
its beginning from the river Rho-
danus (Rhone); it is contained by
the Garumna River, Ocean, and
the borders of the Belgae; the river
Rhenus (Rhine) reaches even from
the Sequani and Helvetii; and it
turns towards the north. The Belgae
arise from the furthest borders of
Gaul, reaching to the lower part of
the Rhenus, and look towards the
north and the rising sun. Aquitania
stretches from the Garumna River
to the Pyrenes mountains and that
part of the Ocean which spans
towards Hispania; it looks between
the setting of the sun and the north.

Mela's Iberia is distinguished by three names: "Part of it [Hispania] is called
Tarraconensis, part Baetica, and part Lusitania." Mela's triple anaphora of
"part" (*pars*) arithmetically duplicates Caesar's "three parts" (*partes tres*). Like
Mela's Hispania, Caesar's Gaul is occupied by three distinct peoples: *Belgae*,
Aquitani, and those "who are called Celts in their own language, but Gauls in
our language." Mela's three Iberian provinces are bordered by three water ways
with four names (*Mari Nostro, Oceano, fluvius Anas,* and *Atlanticum*) in parallel with Caesar's three rivers plus Ocean (*Rhodano, Garumna, Oceano,* and
flumen Rhenum) which separate each of the Gallic provinces. Caesar describes
Gaul as partitioned by three riverine borders together with Ocean. To duplicate
Caesar's watery borders numerically, Mela gives one river, the Anas, and two
larger bodies of water, the Mediterranean and the Atlantic, the latter of which is
cited twice by different names (*Oceanus* and *Atlanticum*).

Caesar highlights each of his Gallic tribes as they are of concern to his readers or of threat to Rome: the Belgae are the bravest (*fortissimi*), as being most
remote from Rome, and engaged in constant warfare with the Germans across
the Rhine; likewise, the Helvetians are braver than the other Celts owing to
their daily military drills. For Mela, the interest of Hispania is not as much in

its military aspects as in its wealth and distinction: Mela tell us, "Hispania is so fertile" that flax, at least, still grows in arid regions. As he often does, Mela lists the most distinguished cities (*clarissimae*), his superlative adjective recalling in tone Caesar's superlatively brave Belgae (*fortissimi*).

Caesar's Belgae share their northeastern border with the Rhine River, "looking toward" (*spectant*) the north (*septentrionem*). Mela repeats Caesar's *spectat* to explain the northern extent (*septentrionem*) of Tarraconensis on the Atlantic. Caesar's Gaul "touches on" (*attingit*) the Rhine River (*flumen Rhenum*, direct object) at the territory of Sequani and Helvetians (expressed with a prepositional phrase: *etiam ab Sequanis et Helvetiis*). Mela inverts Caesar's expression, employing the same verb with a different prefix (*contingens*) to explain that Terraconensis borders two provinces (*Baeticam Lusitaniamque*, direct objects) along the Mediterranean (expressed in the ablative, *Mari Nostro*). Caesar's territorial direct objects are inverted to a prepositional phrase by Mela, whereas Caesar's watery prepositional phrase becomes a direct object in Mela. Finally, Caesar delimits the territory of the Gauls with the verb "to obtain" (*obtinere*). Mela employs the same prefix *ob-*, attached to a different root word to delimit the territory of Lusitania (*obiecta*). In both authors, the prefix *ob* emphasizes territorial limits and edges.

For Caesar, geography and topography are tactical concerns, and his descriptions are vivid and detailed. Both Caesar and Mela are precise in explaining the size and arrangement of each of their three regions, and they use similar language to do so. Mela's tripartite Hispania serves two significant aims: it becomes a microcosm of the entire world, which Mela envisages in his geographic overview as three landmasses, separated by waterways, and echoing geographers from Anaximander onward; and it links our Iberian author to Rome's famous dictator.

Mela treats the territories of the Gauls twice. In the second book, he suggests that Gaul is divided into two parts plus another (third) part (2.74: *Gallia Lemanno lacu et Cebennicis montibus in duo latera divisa, atque altero Tuscum pelagus adtingens altero Oceanum*). Later we read that the peoples of Gallia Comata are distinguished by three names (3.20: *populorum tria summa nomina sunt*—the Aquitani, Celts, and Belgae), and that the races are separated by great bodies of water (Garonne, Seine, and Rhine). Mela repeats the structure of Caesar's map by including Caesar's three rivers in the same order, as well as the dictator's mountain range, the Pyrenees. Mela is self-consciously creating verbal echoes that invite comparison with Caesar. But Mela's Gaul is not Caesar's. The Iberian author provides a fresh account of the geography of Gaul.

Why, we may ask, does Mela evoke Caesar? We know that, unlike other equestrian writers, there is little to suggest that Mela pursued a political career. Rome's more technical geographers, Caesar, Agrippa, and Pliny the Elder, each had extensive firsthand experience of Roman provincial administration which informed their cartographic initiatives, and, like Caesar, both Agrippa and Pliny the Elder aimed to elevate Roman geographical accomplishment in their carto-graphical works and to promote Roman accomplishments abroad, thus exalting Roman culture and virtue. Mela likely lacked this direct experience of gover-nance, and he consequently imitated Caesar to heighten the authority of his own treatise. Mela's work in geography was hardly practical. No magistrate would ever have consulted Mela in any serious way. He was a court geographer, not a magistrate in the provincial field, but he tapped into magisterial authority with subtle allusions to Caesar. In imitating Caesar, Mela heightened the authority of his treatise as an "official" *Roman* geography.

Conclusion

Our Tingenteran author emphasizes his cultural *Romanitas* both explicitly and obliquely. His text reflects Roman concerns and Roman prejudices, as well as unique equestrian and Silver Age interests, including *paradoxa* and technical knowledge. The same can be said for Pliny the Elder. Mela succeeds in pro-ducing something uniquely Roman: a Roman geography written for a strictly Roman audience, reflecting the aesthetic tastes and intellectual interests of the times in which it was penned, and composed from the point of view of a Roman intellectual whose geographical origins are provincial.

Mela strives to establish his identity as culturally Roman, as no doubt he was, but he is still mindful and proud of his provincial Iberian origins. On the one hand, he attempts to compensate for the geographical accident of his birth, with his subtle allusions to Augustus's legacy and Rome's Trojan heritage. On the other hand, he celebrates and showcases his homeland. He promotes the myth of Hercules, whose cult, especially as Gaditanus, was particularly prominent near Gades and Tingentera. Mela emphasizes Phoenician history and settle-ments, and he elevates Pompey (who had campaigned extensively in Iberia) over Augustus's ancestor Julius Caesar (who had merely served a year there as quastor in 69 BC). Finally and significantly, Mela's organizing principle is at the center of this cultural self-awareness. Mela's decision to describe the Medi-terranean world by working counter-clockwise from Mauretania around *Mare Nostrum* in fact allows him to place Tingentera near the center of the treatise. Had he followed the conventional clockwise circuit around the Mediterranean, Tingentera would have been quickly and anticlimactically described, passed over, and forgotten. Instead, Mela deliberately brings his hometown to the cen-ter, framed by the Pillars of Hercules with which the account begins and ends, in an extended hyperbaton.

CHAPTER 5

Mutuo metu aut montibus: Mapping Environmental Determinism in the *Germania* of Tacitus

Molly Ayn Jones-Lewis

FROM ITS EARLIEST DAYS, geography was the study not only of lands but also of the relationship between land and people. The history of geography, therefore, is also the history of the ways in which one group of people made sense of another and brought their biases and assumptions along with them. These assumptions influenced foreign policy and informed how Romans contacted, annexed, and converted territory into Roman provinces. Over time, theories built for one situation mutated in the hands of new imperialists with their eyes on the horizon of conquest; this paper will explore how one such theory—environmental determinism—made its way through earlier texts to serve Tacitus's needs when crafting the tone of the *Germania*. By understanding the way Tacitus's models and sources use (or reject) environmental determinism to structure their ethno-geography, we can better understand the choices Tacitus made and the ways his educated Roman audience would respond to those choices. This reading clarifies the sometimes subtle ways Tacitus uses geographic theory as an invitation to conquest.

Such careful readings are necessary because Tacitus's *Germania* is a cryptic text whose purpose is not obvious, though certainly its date (98 AD) and topic strongly suggest a connection to Rome's Dacian campaigns of the early-second century under Trajan. But the fact that scholars do not completely agree on whether the *Germania* is pro- or anti-invasion—if, indeed, it is about invading Germania at all—highlights the need for multiple approaches to understanding what Tacitus intends for his audience to take from the text. Notable among recent geographically informed readings of the text is Zoë Tan's "Subversive Geography in Tacitus' Germania," which argues (inter alia) that the text's departures from the standard features of geographical writing can be seen as a covert indictment of imperial failures to conquer and incorporate the region.[1] I agree partly with Tan's reading; Tacitus certainly damns Domitian's efforts in the region with his silence. However, Tacitus's geography is not only serving as

---

1. Zoë Tan, "Subversive Geography in Tacitus' *Germania*" *JRS* 104 (2014): 181–204.

a critique of the past, but as an endorsement of more militant future actions. Where Tacitus invokes tropes common to the genre of geography, he does so in order to advance an argument for future conquest. This paper does not attempt to present *the* definitive reading of the *Germania,* but offers *a* reading based on the choices of rhetorical strategies Tacitus seems to have made when consulting his sources.

Environmental determinism is the notion that the physical environment directs and drives the way in which inhabitants of that region grow and develop.[2] It is not an idea for which the Greek and Roman world can take credit; to the contrary, versions of it are found in the independently developed geographical thought of Han China.[3] It is not a theory without some measure of truth: the environment does indeed influence the sorts of traits that are dominant in a region and the resources available to the cultures that grow out of the area. Like most ideas built on a degree of truth, the problem is in the "determinism" part of the idea, which distorts an observation that environment is *a* factor into a belief that environment is the *primary* factor responsible for human difference. Although this idea did not have a name in the premodern world, it had powerful adherents and was widely accepted as true.

Herodotus of Halikarnassos is, perhaps, the first in the Classical tradition to present environmental determinism and apply it widely in service to his ambitions of understanding the world. The theory is most clearly articulated in Book 9, when Cyrus observes that, "from soft places, soft men tend to arise" (*Hist.* 9.122.3). In Greek ethnography, this relationship between place and inhabitant was used to explain human cultural and physical variation through environmental factors. The Hippocratic tradition added the approving stamp of medical theory in the treatise *Airs, Waters, Places* by arguing that there could be such a thing as a place that is *too* favorable, whose land coddles the inhabitants in such a way as to make them morally flabby and unequal to the alleged rigors

---

2. For an overview of the development of theories explaining human difference, see Rebecca Futo Kennedy, "Airs, Waters, Metals, Earth: People and Land in Archaic and Classical Greek Thought," in *The Routledge Handbook of Identity and the Environment in the Classical and Medieval Worlds,* ed. R. F. Kennedy and M. Jones-Lewis (London: Routledge, 2015), 390–412. Other recent considerations of the topic of identity, including environmentally deterministic theories, include Benjamin Isaac, *The Invention of Racism in Classical Antiquity* (Princeton: Princton University Press, 2013), and Erich S. Gruen, *Rethinking the Other in Antiquity* (Princeton: Princeton University Press, 2012).

3. Shao-yun Yang, "'Their Lands are Peripheral and Their *qi* is Blocked Up': The Uses of Environmental Determinism in Han (206 BC–220 AD) and Tang (618–907 BC): Chinese Interpretations of the 'Barbarians'," in *The Routledge Handbook of Identity and the Environment in the Classical and Medieval Worlds,* ed. R. F. Kennedy and M. Jones-Lewis (London: Routledge, 2015), 390–412.

of political independence.[4] Thus embedded deeply into the most influential of early geographers, the idea could not help but become part of the geographical discipline. Herodotus, by putting the words in Cyrus's Persian mouth, presents the idea as a universal one that transcends the immediate context to become a general principle applicable to any people who could be viewed as strong men in a harsh land. But here, too, is one of the troublesome features of this idea because it defines superiority by the lack of refinement and creature comforts.

This concept of the corrupting influence of comfort on the physical and political body is of particular relevance to the way in which Greeks and, later, Romans interacted with the region they called Germania. Germania on its surface ticked every box in the Greco-Roman rubric for lands productive of the strong-man ideal, or *virtus*.[5] Remote from the comforts of aqueducts and bath houses, seemingly content to go without written literature, relatively unstratified in its socioeconomic hierarchies, frigid and wild in its lands, Germania provided a direct challenge to Roman notions of their own superiority. The resulting discomfort can be seen in the way that Germania's hopeful conquerors wrote about the land; Julius Caesar uses German *virtus* to structure the geography in the opening passage of his *Gallic War*. Germania posed a problem for Romans who embraced environmental determinism, and the task of reinterpreting this problem shapes much of the writing about the area, as well as the military and diplomatic policy in the North.

If environment is such a powerful force in shaping human bodies and culture, then the prospect of immigration and expansion—let alone conquest and empire—becomes a riskier business for the conquerors. A more nuanced approach was needed, and one that could be used to lend scientific support to a Roman foreign policy that valued expansion and integration. Such a pragmatic

---

4. περὶ δὲ τῆς ἀθυμίης τῶν ἀνθρώπων καὶ τῆς ἀνανδρείης ... τὰ ἤθεα αἱ ὧραι αἴτιαι μάλιστα. ... διὰ ταύτας ἐμοὶ δοκεῖ τὰς προφάσιας ἄναλκες εἶναι τὸ γένος τὸ Ἀσιηνὸν καὶ προσέτι διὰ τοὺς νόμους. τῆς γὰρ Ἀσίης τὰ πολλὰ βασιλεύεται.

But concerning the lack of courage and unmanliness ... our climates are most to blame. Because of this, the Asian *genos* seems to me by this logic to be feeble. And their *nomoi* contribute to this, for the majority of Asia is ruled by a king (*Airs, Waters, Places* 16).

While this treatise was written to fit the political realities of its time, its influence extended well past the context. See Jacques Jouanna's "Water, Health, and Disease in the Hippocratic Treatise *Airs, Waters, Places*," in *Greek Medicine from Hippocrates to Galen*, ed. Philip van der Eijk and trans. Neil Alles (Leiden: Brill, 2012), 155–72.

5. I will be distinguishing between the Roman construct of Germania and the modern nation-state Germany by referring to the ancient region as Germania. I will also refer to the people of Germania as Germani, rather than Germans, for the same reason. For a fuller discussion of the distinction between Roman Germania and Germany (as well as a different approach to nomenclature), see Christopher B. Krebs, *A Most Dangerous Book: Tacitus's* Germania *From the Roman Empire to the Third Reich* (New York: Norton, 2012), 42–44.

revision is taking place in Tacitus's *Germania*, where the logic of environmental determinism is being used to imply rather than explicitly argue that Germania and Rome need each other, and furthermore that Germania will be better off under Roman control, not because of malleable political conditions, but because Romans have natural qualities that Germani lack. Romans must save the Germani from themselves, and Romans will be improved by their contact with the *virtus* of the German people and the environment that spawned it.

Understanding how Romans grappled with this idea—environmental determinism as it relates to Germania—is the key also to understanding Roman foreign policy in the region. But it has limits too, for it only tells us about Roman assumptions and blindspots. Who the inhabitants of Germania actually were and how they felt about Roman fixations on *virtus* and weather is generally unrecoverable, no matter how one tries to use the archaeological record as a counter-narrative to that of the Greeks and Romans. The attempt to do so is a worthy and important endeavor, but beyond the scope of this chapter.[6] Instead, the task here is to understand the theories influencing Tacitus, as well as the choices available to him when working with sources and predecessors to craft his monograph.

Caesar and the Germani

The relationship between geographical theory and Roman foreign policy in Germania begins with Julius Caesar, although he almost certainly did not invent the ethnym and is not particularly innovative in how he discusses geography. Had he been introducing a new term, surely he would have gone to more pains in explaining what was meant by Germania/Germani when first using the word. What he does introduce, however, is a subtly weaponized version of environmental determinism that frames his war and Romano-Germanic conflict as the inevitable result of environmentally fueled northern aggression. Caesar is also notable as the earliest surviving author to differentiate "Germania" from the Celts of western Europe, popularizing its use, if not inventing it (*Gallic War* 1.1).[7] Indeed, the question of who invented the term "Germania" is an important

---

6. This approach is part of a larger trend in postcolonialist readings of ethnography and geography. For a survey of such readings of Tacitus, see Nancy Shumate's "Postcolonial Approaches to Tacitus," in *A Companion to Tacitus*, ed. Victoria Emma Págan (Chicester: Wiley-Blackwell, 2012), 476–503.

7. This paper uses both Greek and Roman sources, which introduced a problem of whether to adhere to Latin spelling or Greek transliteration. Because this is essentially a paper about Roman attitudes, I have chosen to use the Roman spelling unless Greek conventions better fit the immediate context.

one for the topic at hand, because the invention of a name for a region is itself a creative process in which the geographer organizes the world to fit an agenda.

There is a general consensus that the most likely candidate for introducing, if not inventing, the term Γερμάνοι is Poseidonios, a geographer active in the late-second and early-first centuries BC, whose *Histories* were likely available to Caesar.[8] The details beyond that become much more complicated, both to modern observers and to the sources themselves.[9] The resemblance between the term "Germani" and the Latin word *germanus* (genuine) did not escape Strabo (*Geogr.* 7.1.2), who implies that it is a Roman label that does not represent a great a difference between the people on either side of the Rhine. A little over a century later, Tacitus still approaches the term with caution and explains it as a local tribal name that over time became a name adopted by the entirety of the region (*Germania* 2.3). Still other authors minimize or ignore the "Germani," preferring to preserve the Greek tradition of labeling all northwestern Europeans as Κέλτοι. Whether Poseidonios was among those Greek authors who considered Γερμάνοι a subgroup of Κέλτοι is an open question; it is likely that the wide application of the term to Germania was a Roman adaptation that by Caesar's time had become a convenient boundary.

Be that as it may, the term Germani(a) was certainly codified in Roman minds when Caesar chose to divide his theater of war into tripartite Gaul with a separate Germania across the Rhine.[10] One can, and should, still question whether this was Caesar's deliberate choice, or whether Caesar merely reflects the habits of Roman elites of the first century BC. After all, the lands east of the Rhine and north of the Danube had been of great concern to Rome during the time of Marius, Caesar's uncle, when the Cimbri and Teutones made their incursions into Roman territory. Almost certainly Caesar wants his readers to identify his new Gallic venture with that of his ancestor: if "Germania" had been a geographical term used in Rome during Marius's campaigns, Caesar's choice to invoke it could have great effect. There is, however, a benefit to the choice of terminology that can be more easily read in Caesar's account: he gives a religious border tactical significance.[11] Benjamin Isaac observes that Caesar's division demarcates a

---

8. For Poseidonios's work and life, see Daryn Lehoux, "Poseidonios of Apameia (ca. 110– ca. 51 BC)," *EANS* 691–92. For further commentary about Poseidonios and the Germani, see Ian G. Kidd's commentary on F73 in *Posidonius*, vol. 2, *The Commentary* (Cambridge: Cambridge University Press, 1988). See also Gruen, *Rethinking the Other* 141–42 and Malcolm Todd, *The Early Germans* (Malden, UK: Blackwell, 2004), 1–3.

9. For a fuller discussion of the debates surrounding the introduction and use of "Germani" see Rives's excellent commentary: Tacitus, *Germania* ed. J. B. Rives (Oxford: Clarendon, 1999), 21–27.

10. Interestingly, Caesar demotes the Celts to one of three regions in Gaul, quite the comedown from all Europe: *Gallic War* 1.1.

11. Ellen O'Gorman thoroughly explains the connections between mapping, writing, and conquest. See her "No Place Like Rome: Identity and Difference in the *Germania* of Tacitus," in *Tacitus*, ed. Rhiannon Ash (London: Bloomsbury, 2006), 95–118.

religious boundary as much as it does a geographical one: the Druids described by Caesar were not only a unifying force of belief and ritual for a politically and linguistically diverse region, but they were also a common judicial body whose authority to ban any devotee from the sacrifices could significantly influence secular politics when directed against any Gaulish noble.[12] These Druids' roots in Britannia presented a further challenge to Roman expansion in the form of a group with the potential for uniting Gaulic resistance from a power base far out of Rome's immediate military reach. In this light, Caesar's choice to define Germania has a practical end that can be seen playing out in Rome's military policies of the next two centuries: if the final conquest of Gaul can be found only in the elimination of the Druids, then lands in which the Druids have influence must be conquered before turning to lands beyond their unifying influence. Germania has no Druids (*Gallic War* 6.10). Germania can wait until Britannia is pacified.

Nevertheless, Isaac also sees in Caesar a departure from the more heavy-handed environmental determinism of his predecessors, and this is true to a certain extent.[13] Caesar spends little time theorizing about the effects of cold dampness on his enemies and avoids, in large part, obvious references to and usage of Greek environmental theory, giving instead the impression of original autopsy informed only rarely by scholarly resources. But to see in Caesar a narrator resistant to or uninformed by the theoretical tradition of climate and human difference is to miss a key element of his rhetorical strategy and to give in to a self-presentation that deemphasizes Caesar's command of Greek learning.[14] The future dictator had a light hand when he invoked elite assumptions about the North, but invoke them he did.

Indeed, Caesar drafts his opening map with lines drawn in Roman cultural anxiety:

> Gallia est omnis divisa in partes tres, quarum unam incolunt Belgae, aliam Aquitani, tertiam qui ipsorum lingua Celtae, nostra Galli appellantur. Hi omnes lingua, institutis, legibus inter se differunt. Gallos ab Aquitanis Garumna flumen, a Belgis Matrona et Sequana dividit. Horum omnium fortissimi sunt Belgae, propterea quod a cultu atque humanitate provinciae longissime absunt, minimeque ad eos mercatores saepe commeant atque ea quae ad effeminandos animos pertinent important, proximique sunt Germanis, qui trans Rhenum incolunt, quibuscum continenter bellum

---

12. Isaac, *Invention of Racism*, 421–25.

13. Isaac, *Invention of Racism*, 413–14.

14. Isaac, *Invention of Racism*, esp. 413–14, sees Caesar as somewhat independent from environmental theories in preference of cultural factors. Although Caesar's acceptance of environmental determinism is certainly nuanced, one cannot escape the fact that he opens his work with an argument based on proximity to the Mediterranean as responsible for relative *virtus*.

gerunt. Qua de causa Helvetii quoque reliquos Gallos virtute praecedunt, quod fere cotidianis proeliis cum Germanis contendunt, cum aut suis fini- bus eos prohibent aut ipsi in eorum finibus bellum gerunt.

The entirety of Gaul is divided into three parts, one of which the Belgae inhabit, the other the Aquitani, the third those who are called Celtae in their own language and Galli in ours. All these regions differ amongst themselves in language, government, and laws. The Garumna river [Garonne] divides Gauls from Aquitanians, the Matrona [Maronne] and Sequana [Seine] divide Gauls from Belgae. The strongest of all of these are the Belgae due to the fact that they are the farthest distant from the sophistication and civilization of the Province, and merchants come least frequently to them and bring the things that tend to feminize minds, and they are closest to the Germani, who live across the Rhine, and with whom they ceaselessly wage war. For this reason the Helvetii also outstrip the rest of the Gauls in *virtus*, because they skirmish in almost daily battles with the Germani, when they either keep [the Germani] out of their bor- ders or themselves raid within the borders of [the Germani]. (Caesar, *Gallic War* 1.1)[15]

The first thing Julius Caesar tells his readers—the Roman public *and* the senate whom he wishes to convince to support his war of thinly veiled aggression— is that the *virtus* of the three parts of Gaul is directly related to the proximity to the *Germani qui trans Rhenum flumen habitant*. While it is not climate so much as access to trade that Caesar names as the proximate cause for losing *virtus* (*quae ad effeminandos animos pertinet*, "the things that tend to feminize minds"), the tradition of relating harshness in the environment to *virtus* is not irrelevant to either Caesar or his audience. Caesar can be so minimalistic in his introductory ethnogeographical primer *because* he is invoking a familiar ethno- graphic trope in service of his goals of inflating the trans-Rhenic menace and emphasizing the immediate peril posed to Rome. As others have noted, Caesar's invocation of German menace recalls, in a way that must have been obvious to Roman readers, the invasion of Italy by the Cimbri and Teutones, which only Marius was able to stop during his campaigns of 104–101 BC. It is, therefore, hardly surprising that Caesar launches directly from a hastily sketched map of the region into the migration of the Helvetii; the parallels would be clear to a Roman

15. The chapter returns to mapping in the final (omitted) three lines and sketched the borders of these three regions along the boundaries of mountains, rivers, and ocean. Caesar then launches into the story of Orgetorix and the Helvetian migration out of their cramped Rhine-adjacent territory onto the edge of territory controlled by the Allobroges, friends of Rome whom Caesar claims to aid when he initiates hostilities.

reader. This accounts for the awkward asymmetry introduced when he turns from a discussion of how the Belgae, being farthest from the Romanized lands of Gallia Transalpina and closest to Germania, are the strongest (*fortissimi*) inhabitants of Gaul to a discussion of the martial superiority of the Helvetii (*Helvetii quoque reliquos Gallos virtute praecedunt*). Which is it? Are the Belgae strongest, or are the Helvetii, who are located directly opposite the Alps from Rome and are, therefore, aberrations from a model locating maximum *virtus* farthest from the Mediterranean? Caesar has it both ways with his *quoque*, citing improbable, near daily battles with Germani (*fere cotidianis proeliis*) as the reason for the paradigm breaking *virtus* of the Helvetii. The implication is that the people across the Rhine are so filled with *virtus* that if one fights them often enough, one's own *virtus* is increased by mere contact. Caesar certainly had ambitious goals, and this literary strategy, forged to further his personal agenda, was persuasive enough to set the location of Rome's aspirational boundaries for generations to come. Indeed, it persisted after Caesar's death and was directly responsible for a lengthy, costly, and ultimately fruitless effort to expand beyond the Rhine and Danube. It is this goal that Tacitus is still eyeing a century and a half later.

But what his geographical model implies about Rome and its readiness to face the crisis that Caesar presents here is more subtle. If Germania and the Rhine are the locus for all things rugged, harsh, and unrelentingly bellicose, then Rome occupies an opposing pole. Caesar is careful not to claim outright that Rome is soft, effeminate, and a corrupting influence. Caesar is not attempting political suicide. However, the *mercatores* whom he invokes as being responsible for the *cultus et humanitas* of Gallia Transalpina can be understood as a consequence of trade contacts with Rome, or at least the Mediterranean region in which Rome was then dominant. He definitely invokes the trope of the corrupting sea.[16] Caesar's source of *humanitas* falls short of Rome itself; he names the Roman province Gallia Cisalpina (*Provincia*) as the epicenter of mercantile *luxuria*. This is not just a map about what Rome wants, but a map of what Caesar suggests Rome needs: an infusion of *virtus* that can be gained only from ongoing combat with the unblunted ferocity of the North. Germania is not simply a source of danger but a source of self-improvement for its foes.

Caesar in Germania

This tactic proved successful enough that Caesar further elaborated on it in subsequent books of the *Gallic War*, where he pauses in his war stories to fill

---

16. For a substantial discussion of this trope, see Peregrine Horden and Nicholas Purcell, *The Corrupting Sea: A Study of Mediterranean History* (Oxford: Blackwell, 2000).

in the lines drawn in his opening chapter, first in a brief aside about the subpar cattle, militaristic focus, and physical size of the Suebi in *Gallic War* 4.1–2,[17] and at greater length in a more purposeful ethnography culminating in a memorable tour of the Hercynian forest at 6.21–28 drawn from Eratosthenes.[18] Although the environmental cues are subtle, they are nonetheless present in observations about the lack of clothing relative to the cold (4.1; 6.21). Less obvious to a modern eye is this observation at the end of 6.21.3–5:

Qui diutissime impuberes permanserunt, maximam inter suos ferunt laudem: hoc ali staturam, ali vires nervosque confirmari putant. Intra annum vero vicesimum feminae notitiam habuisse in turpissimis habent rebus; cuius rei nulla est occultatio, quod et promiscue in fluminibus perluuntur et pellibus aut parvis renonum tegimentis utuntur magna corporis parte nuda.

Those [young folk] who have remained chaste the longest, have the greatest respect among their people: by this, some think that height is established, others strength and endurance. Indeed, they consider it among the most shameful things to have had knowledge of a woman before one's twentieth year; of which fact there is no concealment, because they also bathe together in rivers and use pelts or small coverings of deer-hide as covering, leaving a great part of the body nude.

How would group bathing in rivers betray whether or not a young man has lost his virginity? For that logic, we must turn to the Hippocratic treatises *On the Seed* 1–2 and *On the Child* 20, which explain that seminal fluid drives the growth of body and pubic hair.[19] The more one emits seminal fluid from the head to the groin, the more hair grows on the body. One can imagine that this belief made many a young man's adolescence that much more awkward. It is possible that Caesar was familiar with this sort of medical theory, but it is perhaps more likely that he reflects a source steeped in the Greek intellectual tradition that generated such beliefs: Poseidonios, again, is the most likely suspect. Be that as it may, invoking such a specifically Greco-Roman idea,[20] and then assuming that

---

17. This passage reads like an embryonic form of Tacitus's more embellished observations on German cattle at *Germania* 5, suggesting a common source (Poseidonios seems most likely), direct reference, or a combination of the two seasoned with a liberal dash of Tacitean sneer.

18. Caesar provides a rare explicit reference to Eratosthenes at 6.21.

19. *On the Seed* is also known as *On Generation*. Interested readers will be well served by Iain M. Lonie's *The Hippocratic Treatises "On Generation," "On the Nature of the Child," "Diseases IV": A Commentary* (Berlin: de Gruyter, 1981).

20. The degree to which Greek medical theory had gained acceptance in Caesar's Rome is still a matter of debate, but in this context, it seems less accurate to class the elite medical culture that formed Caesar's assumptions about the human body as "Greek." Roman patronage had already

it is shared by the Germani under discussion is a significant choice on Caesar's part, whether conscious or not. In one stroke, he validates the inherent rightness of Greco-Roman science while also reminding the audience of other related beliefs about cold and its effect on the human body and mind. Caesar's overt explanations for Germanic aggression, size, and strength are based mostly in custom, but the covert implications of environmental determinism are still present, lurking beneath the harsh customs and observations about material culture in a foundation that Tacitus will embellish and bring back to the foreground.

Strabo, Pliny, and Alternative Germaniae

Although Strabo was probably not a source for Tacitus (his manuscript was not widely available until the second century AD),[21] his text offers invaluable comparanda for Caesar's use of Poseidonios, as well as an approach to human diversity that contrasts greatly with the bellicose tone of both Caesar and Tacitus. Although environmental determinism is not absent from the text, Strabo uses such ideas gently when it comes to Gaul and Germania. One must take care when supplying explanations for his choice of tone, but even our scant knowledge of his life provides some suggestions. A native of Amaseia in Pontos born sometime in the 60s BC, he had a Roman cognomen (likely gained through patronage, possibly by Aelius Gallus)[22] and spent (perhaps) considerable time in Rome before moving on to Alexandria in Egypt as part of Gallus's staff. His *Geography* betrays an abiding interest in mining and fisheries; perhaps his duties included some form of inspection tours. What differentiates him from the other authors before us is his immigrant status and a perspective informed by exposure to many cultures while living the majority of his life as a proud Pontic expatriate. Tourism hardly immunizes an author from generalizations and stereotypes: Herodotus is in many ways a parent to the strong vein of environmental determinism informing subsequent Greek ethnography. But Strabo's personal experience as both conquered and a conqueror's aide may very well provide us with insights into why he chose to approach the North in a way that feels Romanocentric in form but more ecumenical in tone.[23]

---

entangled itself sufficiently, into the evolving theories of Greek medicine in the first century BC that calling Caesar's medical beliefs "Greco-Roman" seems to best reflect the messy reality. For Roman reception of Greek medicine, see Vivian Nutton, *Ancient Medicine,* 2nd ed. (London: Routledge, 2013), 160–73.

21. For the life of Strabo, see Duane W. Roller, *The Geography of Strabo: An English Translation with Introduction and Notes* (Cambridge: Cambridge University Press, 2014), 27–39.

22. See Roller *Geography of Strabo*, 2n6.

23. This humanizing approach to discussing the Other is obvious from the very first instance of Germania in Strabo, *Geogr.* 1.17, where he maintains Caesar's (and likely Poseidonios's) division

Strabo is agreeably diligent in informing the audience about his sources; this is fortunate, as it allows us to compare passages and detect Poseidonios's tracks in Caesar's prose, the strongest parallel for which can be found in his discussion of the peoples of Gaul. That parallel is itself telling. Compare Caesar's description of the Belgae at *Gallic War* I.I, quoted above, to the following:

εἰσὶ μὲν οὖν μαχηταὶ πάντες τῇ φύσει, κρείττους δ᾽ ἱππόται ἢ πεζοί, καὶ ἔστι Ῥωμαίοις τῆς ἱππείας ἡ ἀρίστη παρὰ τούτων. ἀεὶ δὲ οἱ προσβορρό-τεροι καὶ παρωκεανῖται μαχιμώτεροι, τούτων δὲ τοὺς Βέλγας ἀρίστους φασίν, (εἰς πεντεκαίδεκα ἔθνη διῃρημένους τὰ μεταξὺ τοῦ Ῥήνου καὶ τοῦ Λίγηρος παροικοῦντα τὸν Ὠκεανόν), ὥστε μόνους ἀντέχειν πρὸς τὴν τῶν Γερμανῶν ἔφοδον Κίμβρων καὶ Τευτόνων· αὐτῶν δὲ τῶν Βελγῶν Βελλοάκους ἀρίστους φασί, μετὰ δὲ τούτους Σουεσσίωνας.

τῆς δὲ πολυανθρωπίας σημεῖον· εἰς γὰρ τριάκοντα μυριάδας ἐξετάζε-σθαί φασι τῶν Βελγῶν πρότερον τῶν δυναμένων φέρειν ὅπλα· εἴρηται δὲ καὶ τὸ τῶν Ἐλουηττίων πλῆθος καὶ τὸ τῶν Ἀρουέρνων καὶ τὸ τῶν συμ-μάχων. ἐξ ὧν ἡ πολυανθρωπία φαίνεται καὶ, ὅπερ εἶπον ἡ τῶν γυναικῶν ἀρετὴ πρὸς τὸ τίκτειν καὶ ἐκτρέφειν τοὺς παῖδας.[24]

They are all fighters by nature and are better as cavalry than infantry: the best Roman cavalry comes from them. Those more toward the North and along the ocean are always more warlike. They say that the Belgians are the bravest of these, and they are divided into fifteen peoples living along the ocean between the Rhenos and the Liger, and thus were the only ones who could hold out against the German invasion by the Kimbrians and Teutonians. They say that the Bellovacians are the bravest of the Belgians, and after them the Souessionians.

This is an indicator of their large population: those who have examined this carefully say that formerly there were up to 300,000 Belgians able to bear arms. The number of the Elvettians, Arvernians, and their allies has already been told, from which the size of the population can be shown, and also, as I have said [4.1.2], the excellence of the women in regard to the bearing and nursing of their children. (Strabo, *Geogr.* 4.4.3 [Roller])

---

of Celt and German in the same breath that he mentions both people's use of topography to resist Roman invasion. It is telling that Strabo focuses that observation from the point of view of the invaded; one is tempted to compare similar tactics employed in the Mithradatic Wars by the locals to stymie Pompey's advancing forces with poisoned honey and other local inconveniences (Strabo, *Geogr.* 12.3.18). Interestingly, Benjamin Isaac sees Strabo as more strongly influenced by environ-mental theories and stereotypes than Caesar, and he reads in Strabo's *Geography* less sympathy for Gaul and Germania (Isaac, *Invention of Racism*, 429–31).

24. I use Stefan Radt's edition for my text here and elsewhere: *Strabons Geographika*, vols. 1 and 2 (Göttingen: Vandenhoeck & Ruprecht, 2002–2011).

There are no southern *mercatores* lurking about with their supplies *ad effemi-nandos animos* here. Rather, Strabo's claims about relative bravery rely on a *specific* instance of resistance to German invasion and immigration instead of allegations of regular and ongoing altercations with *Germana virtus*. It is interesting that he, and not Caesar, brings up the Cimbri and Teutones. Caesar doubtless did not need to be so specific when his audience was well aware of the parallel, and that does account for some of the choices. The sources Strabo cited here regarding the Belgae include Caesar himself, Pytheas, and Poseidonios. Since we can compare Caesar, it is possible that the tidbit about the Cimbri and Teutones comes from Poseidonios and was omitted in Caesar's version or that Strabo made the implication plainer for a broader audience. In either case, Strabo's account decouples German aggression from German nature in a way that parallels his earlier brief description of the local Gaulish/German character:[25]

τὸ δὲ σύμπαν φῦλον ὃ νῦν Γαλλικόν τε καὶ Γαλατικὸν καλοῦσιν, ἀρει-μάνιόν ἐστι καὶ θυμικόν τε καὶ ταχὺ πρὸς μάχην, ἄλλως δὲ ἁπλοῦν καὶ οὐ κακόηθες. διὰ δὲ τοῦτο ἐρεθισθέντες μὲν ἀθρόοι συνίασι πρὸς τοὺς ἀγῶνας καὶ φανερῶς καὶ οὐ μετὰ περισκέψεως, ὥστε καὶ εὐμεταχείριστοι γίνονται τοῖς καταστρατηγεῖν ἐθέλουσι· καὶ γὰρ ὅτε βούλεται καὶ ὅπου καὶ ἀφ᾽ ἧς ἔτυχε προφάσεως παροξύνας τις αὐτοὺς ἑτοίμους ἔσχε πρὸς τὸν κίνδυνον, πλὴν βίας καὶ τόλμης οὐδὲν ἔχοντας τὸ συναγωνιζόμενον. παραπεισθέντες δὲ εὐμαρῶς ἐνδιδόασι πρὸς τὸ χρήσιμον, ὥστε καὶ παι-δείας ἅπτεσθαι καὶ λόγων. τῆς δὲ βίας τὸ μὲν ἐκ τῶν σωμάτων ἐστὶ μεγά-λων ὄντων, τὸ δ᾽ ἐκ τοῦ πλήθους· συνίασι δὲ κατὰ πλῆθος ῥαδίως διὰ τὸ ἁπλοῦν καὶ αὐθέκαστον, συναγανακτούντων τοῖς ἀδικεῖσθαι δοκοῦσιν ἀεὶ τῶν πλησίον. νυνὶ μὲν οὖν ἐν εἰρήνῃ πάντες εἰσὶ δεδουλωμένοι καὶ ζῶντες κατὰ τὰ προστάγματα τῶν ἑλόντων αὐτοὺς Ῥωμαίων, ἀλλ᾽ ἐκ τῶν παλαιῶν χρόνων τοῦτο λαμβάνομεν περὶ αὐτῶν <καὶ> ἐκ τῶν μέχρι νῦν συμμενόντων παρὰ τοῖς Γερμανοῖς νομίμων.

The entire race that is now called Gallic or Galatic is intense about war, and high spirited and quick to fight, but otherwise simple and not malig-nant. Because of this, when roused to fight they come together in a body for the struggle, openly and without consideration, so that those who wish to defeat them by stratagem can do so easily. If one is willing to provoke

---

25. Strabo's accounting for the ethnym *Germani* signals his awareness that the term is novel. Unlike Tacitus (*Germania* 2), he claims that it is a Roman invention based on a Roman assumption that the *Germani* are the "genuine" (*germanus*) version of other Keltic peoples. Strabo may well be correct. Strabo, *Geogr.* 7.1.2.

them by whatever means on some pretext, they will readily go into danger, having nothing to assist them except strength and courage. But if persuaded, they easily give in to usefulness, so that they engage in education and language. Their strength comes from the size of their bodies and from their numbers. Because of their simplicity and bluntness, they easily come together in a multitude, always having a common anger with their neighbors whom they think have been wronged. (At present they are all at peace, since they have been <u>enslaved and live according to the dictates of the Romans who captured them</u>, but we are taking the account from ancient times, as well as the customs that still remain today among the Germans. (Strabo, *Geogr.* 4.4.2 [Roller])[26]

Like his predecessors, Strabo admits to cultural overlap in Gaul and Germania, but he avoids Caesar's vivid invocation of Druidic authority (Strabo, *Geogr.* 4.4.3–5, vs. Caesar, *Gallic War* 6.13–14, 16,). Caesar stresses the danger of Druids and their (allegedly) absolute authority, giving the impression of omnipresent violence and dictatorial control, while Strabo permits Bards and Vates to join the Druids in the religious life of Gaul and seems by turns approving of the checks they impose on warfare and amused by the fondness of Rhineland Gauls for jewelry and emotional displays. Caesar wishes his readers to believe that Rome must save Gaul from itself. Strabo, looking at the results of Caesar's policies, seems unconvinced, and his account of Druidic human sacrifice is much less iconic that Caesar's lurid prose in *Gallic War* 6.16. The closest Strabo comes to expressing outright approval of Roman conquest is this passage from *Geogr.* 4.4.5: καὶ τούτων δ' ἔπαυσαν αὐτοὺς Ῥωμαῖοι καὶ τῶν κατὰ τὰς θυσίας καὶ μαντείας ὑπεναντίων τοῖς παρ' ἡμῖν νομίμοις ("The Romans also stopped them from doing these things, the practices contrary to our customs both at sacrifices and at divinations"). Drusus and Germanicus were able to convert the Tauriskians into quiet tribute-paying neighbors to Rome's north, again with implied approval (Strabo, *Geogr.* 4.6.9). Indeed, Strabo has frank words about Rome's methods of suppressing an inclination to bellicosity that avoids judgmental language for the Gauls and, indeed, the Germani whom he uses as a way of accessing an unromanized version of Gaul. He maintains this tone even while quoting passages of Poseidonios that seem to have been quite inflammatory and judgmental:

πρόσεστι δὲ τῆι ἀνοίαι καὶ τὸ βάρβαρον καὶ τὸ ἔκφυλον, ὃ τοῖς προσβόρροις ἔθνεσι παρακολουθεῖ πλεῖστον, τὸ ἀπὸ τῆς μάχης ἀπιόντας τὰς κεφαλὰς τῶν πολεμίων ἐξάπτειν ἐκ τῶν αὐχένων τῶν ἵππων, κομίσαντας

---

26. Strabo goes on to make further comments on the similarities between Gauls and Germani.

δὲ προσπατταλεύειν τοῖς προπυλαίοις. φησὶ γοῦν Ποσειδώνιος αὐτὸς
ἰδεῖν τὴν θέαν ταύτην πολλαχοῦ καὶ τὸ μὲν πρῶτον ἀηθίζεσθαι, μετὰ
δὲ ταῦτα φέρειν πρᾴως διὰ τὴν συνήθειαν. τὰς δὲ τῶν ἐνδόξων κεφαλὰς
κεδροῦντες ἐπεδείκνυον τοῖς ξένοις καὶ οὐδὲ πρὸς ἰσοστάσιον χρυσὸν
ἀπολυτροῦν ἠξίουν.

In addition to their foolishness, there is the barbarity and alienness that
is an attribute of most peoples toward the north, so that when going
away from a battle they fasten the heads of their enemies to the necks
of their horses and when returning [home] they nail them to their gates.
Poseidonios,[27] at any rate, says that he saw this in many places, and at
first found it disgusting but later bore it mildly because of its familiarity.
(Strabo, *Geogr.* 4.4.5 [Roller])

Even this head-hunting is softened in Strabo's account (perhaps following Poseidonios) as an eccentricity to which one can be accustomed. The aggressors in
the Rhineland, as in other parts of Germania, are sometimes other Germani/
Celts but primarily Romans.[28] Strabo is a careful critic of Roman influence in
the region, but he is a critic nonetheless.

Pliny

That the Elder Pliny's *German Wars* was a likely source for Tacitus's monograph
has been an object of scholarly speculation since Ronald Syme.[29] Tacitus's close
association with Pliny's family almost guarantees that he consulted the work.
However, the *German Wars* are lost, and this hampers efforts to determine how
much influence Pliny the Elder had over Tacitus's text. The fragments are sparse
indeed, occupying no more than a handful of pages in the *Fragments of the
Roman Historians*.[30] This leaves us with the less helpful option of comparing
what Pliny says about Germania in the *Naturalis Historia* to Tacitus; the effort
must be attempted with caution, but I will assume that Pliny's opinions about the

27. See Fragment 274 in Kidd's *Posidonius*, vol. 2, *The Commentary*.
28. See especially Strabo, *Geogr.* 4.6.7; 7.1.3 (in which Romans displace Germani in the Rhine
valley); 7.1.4 (which includes a balanced account of the Teutoberg incident, in which the Cherusci
are blamed for treachery, but Germanicus's vengeance is showcased as destructive and sufficient
compensation for that choice).
29. Ronald Syme, *Tacitus* (Oxford: Oxford University Press, 1958), 127.
30. Barbara Levick provides text and commentary for the very few fragments of the *Bella
Germaniae* in Tim J. Cornell, *The Fragments of the Roman Historians* (Oxford: Oxford University
Press, 2013), no. 80.

region and approach to discussing it did not change drastically between writing the *Naturalis Historia* and the *Bella Germaniae*.

What becomes quickly clear in comparing Pliny's geographical descriptions of Germania with Tacitus's are the sharp dissimilarities in phrasing, even as both authors trace the same general landmarks. For the Danube, Pliny (*Nat. hist.* 4.79) and Tacitus (*Germania* 1.1) both trace a course from Mt. Abnoba to its six mouths opening onto the Black Sea, but Pliny includes the names of local peoples where Tacitus glosses over them with *plurimos populos adit* and chooses more vivid verbs for his river's actions (*effusus, erumpat, os hauritur* vs. Pliny's *ortus, adluit, evolvitur*). Pliny's Rhine is even less similar to Tacitus's portayal: *ita appellantur ostia, in quae effusus Rhenus a septentrione in lacus, ab occidente in amnem Mosam se spargit, medio inter haec ore modicum nomini suo custodiens alveum* (*Nat. hist.* 4.101). And Tacitus: *Rhenus, Raeticarum Alpium inaccesso ac praecipiti vertice ortus, modico flexu in occidentem versus septentrionali Oceano miscetur* (*Germania* 1.1). There is one notable exception to this lack of intertextuality between Pliny and Tacitus:

nam et a Germania inmensas insulas non pridem conpertas cognitum habeo.

Because I understand that even off the coast of Germania huge islands not previously known have been discovered. (Pliny, *Nat. hist.* 2.246)

cetera Oceanus ambit, latos sinus et insularum inmensa spatia complectens, nuper cognitis quibusdam gentibus ac regibus, quos bellum aperuit.

The Ocean encircles the rest, embracing wide bays and huge expanses of islands, since certain peoples and kings have been recently discovered, which war revealed. (Tacitus, *Germania* 1.1)

Granted, the words are not so similar as to prove intertext, but *inmensa, insulae,* and *cognitus* in the context of newly discovered northern territory do suggest that Tacitus was either looking at Pliny's text or a common source. This seems, however, to be the extent of Tacitus's clear debt to Pliny. Unless the tone of the *Bella Germaniae* was vastly different from the *Naturalis Historia,* Tacitus seems to have avoided direct references.

This is not to say that Pliny's influence cannot be seen in Tacitus's approach, especially in the way assumptions of environmental determinism underpin Pliny's discussions of human difference and the natural world.[31] He is typical of

---

31. For the full argument and a table of citations see Molly Ayn Jones-Lewis, "Poison: Nature's Argument for the Roman Empire in Pliny the Elder's *Naturalis Historia,*" *CW* 106 (2012): 51–74.

Greek and Roman natural scientists in focusing on cold as the dominant natural force in the North, and firmly classifies Germania as an area emblematic of the pervasive effects of chill and damp.[32] Most of Pliny's claims about Germania's nature are neutral in tone, or occasionally positive.[33] In two striking cases, however, German nature proves particularly hostile to Romans. At 25.20, in one of several anecdotes taken from Germanicus's German campaigns and therefore likely to have been featured in the *Bella Germaniae*, Pliny discusses a spring that, if used as a water source for more than two years, caused the Romans' knees to weaken and their teeth to drop out until an herb called *britannica* was administered. At 16.5 German *natura* attacks Roman forces more literally:

Aliud e silvis miraculum: totam reliquam Germaniam operiunt adduntque frigori umbras, altissimae tamen haud procul supra dictis Chaucis circa duos praecipue lacus. litora ipsa optinent quercus maxima aviditate nascendi, suffossaeque fluctibus aut propulsae flatibus vastas complexu radicum insulas secum auferunt, atque ita libratae stantes navigant, ingentium ramorum armamentis saepe territis classibus nostris, cum velut ex industria fluctibus agerentur in proras stantium noctu, inopesque remedii illae proelium navale adversus arbores inirent.

Another marvel from the forests: they cover all the remaining parts of Germany and add shade to the cold, the highest not far from the aforementioned Chauci especially around the two lakes. Oaks crowd the very shores with their hunger for growing to their peak, and washed out by the waves or pushed out by winds they carry off vast islands of roots with them in their embrace, and standing thus unmoored they sail the waters, and have often terrorized our fleets with the armaments of their huge branches, when they are either driven by the waves as if by some conscious effort into the prows of our ships anchored in the night, and at a loss for a solution [our ships] have joined naval battle against trees.

Pliny then goes on to discuss the significance of the oak crown as a Roman military award, neatly pivoting away from the horror of German lakelands back

32. See especially Pliny, *Nat. hist.* 10.72, in which winter-friendly thrushes populate Germania; 11.33, in which Germania's northerly location produces huge, black-cored honeycomb; 18.149, in which Germania's dampness and the weakness of local seed is blamed for wheat and barley degrading (*degenerat*) into oats; 19.83, in which Germania's chill causes radishes to become huge. Interestingly, most of these are neutral phenomena—the worst effect Pliny attributes to Germania's weather is the "*vitium*" of oats existing.

33. In particular, Pliny praises the large fish of the Danube (*Nat. hist.* 9.45) and German pasture land (17.26).

to Roman valor. However, the point has been made; cold and damp inspire the very trees to acts of military courage, or at least sudden acts of destructive violence. Without deliberately invoking Caesar's claims of German *virtus*, Pliny has referenced and reinforced the stereotype so subtly that it is unclear whether he meant it as a deliberate rhetorical strategy.

That said, Pliny does not single out Germania for any special attention, positive or negative. This is not the case with other locations on Pliny's map or their inhabitants. Greek physicians, for instance, bother him enough to evoke several chapters of warnings and caveats (*Nat. hist.* 29.11–28), but there is no similar section listing the perils of German tribesmen, despite his established interest in both the region and its military history. Indeed, Pontus and its native Psylloi seem to worry him more than does Germania.[34] Here lies another reason to look beyond Pliny for the source of Tacitus's tone.

The model Tacitus does invoke, quite deliberately, is Caesar's. It is the first thing he does, recalling the first words of the *Gallic War* in his own opening lines:

Germania omnis a Gallis Raetisque et Pannoniis Rheno et Danuvio fluminibus, a Sarmatis Dacisque mutuo metu aut montibus separatur: cetera Oceanus ambit, latos sinus et insularum inmensa spatia complectens, nuper cognitis quibusdam gentibus ac regibus, quos bellum aperuit.

All Germania is separated from the Gauls and Raetians and the Pannonians by the Rhine and Danube rivers, and from the Sarmatians and Dacians by either mutual fear or mountains: the Ocean encircles the rest, embracing wide bays and immense stretches of islands, [in which] certain peoples and kings, whom war has exposed, have been recently discovered. (Tacitus, *Germania* 1)[35]

Of course Tacitus also uses Poseidonios (best seen in comparison to Strabo's testimonia), but his closest model in both approach (light on the citations) and tone (bellicose) is the very Caesar who began the parade of *principes* that Tacitus so detests in his larger historical works. Indeed, Caesar gets one of Tacitus's rare

---

34. Pliny is so fixated on the Psylloi of Pontus that he is the main source for much of what we known about them. For the Psylloi, see Molly Ayn Jones-Lewis, "Tribal Identity in the Roman World: The Case of the Psylloi," in *The Routledge Handbook of Identity and the Environment in the Classical and Medieval Worlds*, ed. Rebecca Futo Kennedy and Molly Ayn Jones-Lewis (London: Routledge, 2015), 192–209.

35. Another striking parallel can be seen between *Gallic War* 6.17 and *Germania* 9. The statement, *Deorum maxime Mercurium colunt* ("Of the gods, they especially worship Mercury") opens both passages.

attributions at *Germania* 28. Why that choice of citations? Historical context provides the most likely answer.

Environmental Determinism, Geography, and Desire

In 98 AD, Trajan succeeded the elderly emperor Nerva. In the same year, Tacitus composed and circulated his work about Germania. A conflict between Rome and the Germani across the Danube had been simmering since Domitian's Dacian war, which had ended with a treaty and withdrawal in 88 AD. With the accession of a new emperor—one known for his military prowess—a rematch was brewing that would culminate in the Dacian campaigns of the first decade of the second century. In this political climate, then, Tacitus composed his short ethnography *Germania*, a work that on its surface is little more than a survey of the lands and peoples and that eschews, as Krebs observes,[36] all but the most oblique references to recent military activities but whose subtext cannot help but be imperialist in its intent. But the details of that covert argument hinge on environmental determinism, and I highlight this theory in the text in order to illustrate Tacitus's logic of empire.

Tacitus's environmental determinism is foregrounded in the early chapters of the work, beginning with discussions of German isolation and autochthony and moving into even more explicit discussions of connection between Germani and their land. Section 2 of the *Germania* does not so much articulate environmental determinism as it illustrates its effects: in order to make the best case for a strong connection between the nature of Germani and the nature of Germania, Tacitus emphasizes their isolation and exemption from the colonization that was the norm for other areas of the ancient western world.

> Ipsos Germanos indigenas crediderim minimeque aliarum gentium adventibus et hospitiis mixtos, quia nec terra olim, sed classibus advehebantur qui mutare sedes quaerebant, et inmensus ultra (utque sic dixerim) adversus Oceanus raris ab orbe nostro navibus aditur. <u>Quis porro, praeter periculum horridi et ignoti maris, Asia aut Africa aut Italia relicta Germaniam peteret, informem terris, asperam caelo, tristem cultu adspectuque, nisi si patria sit?</u>

I have been given to believe that the Germani themselves are native, and very little mixed with visitors from other peoples and with guest-friends,

---

36. See Krebs, *A Most Dangerous Book*, 29–55. This reading of Tacitus's geography relies on Tan, "Subversive Geography," 181–204.

because long ago it was not that case that those who wished to change their homeland were carried by land, but rather by sea, and moreover (as I have already said) the vast ocean is approached by only the occasional ships from our sphere. <u>Who, moreover, setting aside the peril of a terrible and unknown sea, after leaving Asia or Africa or Italy behind, would head for Germania, a place shapeless in its lands, harsh in its weather, and gloomy in its cultivation and appearance, unless it were his fatherland?</u> (Tacitus, *Germania* 2)

It is probably appropriate to read this section as a neutral—or even pejorative—observation rather than a compliment; most Romans of the first century accepted origin stories that painted Rome as a place heterogenous at its founding, foregrounding their alloy of Trojan and Italian origins in their public art and literature. The fact that Tacitus uses this statement as a prelude to a vivid description of forbidding and unpleasant geography supports this reading, as does a particularly often quoted line, underlined in passage above. In Tacitus's rhetoric, the population of Germania remains unmixed in an age of improved land travel because nobody *wants* to go there, and not because of some sort of superiority. Strabo's observation (*Geogr.* 7.1.2) that the Romans invented the very name *Germani* due to a belief that Germani were somehow the "genuine" variety of Celts seems pertinent here; Strabo's "genuine" becomes Tacitus's unmixed, homogenous population.[37]

Tacitus moves from isolation into autochthony—that is, the belief that a group of people have been born from the very soil they inhabit.

Celebrant carminibus antiquis, quod unum apud illos memoriae et annalium genus est, Tuistonem deum terra editum. Ei filium Mannum, originem gentis conditoremque.

They proudly claim in ancient songs, which are the one form of memory and annals among them, that their god Tuisto was born from the earth. He had a son named Mannus, the origin of the people and its founder. (*Germania* 2)

While attempts have been made to identify this Tuisto and his son Mannus with later Germanic mythology,[38] the most relevant point here is that Tacitus

---

37. Like Strabo, Tacitus expresses some uncertainty and confusion about the "Germani/a" term later in the second chapter, where he observes that the term *Germani* is a newer one then tentatively suggests some possible derivations for it.

38. See J. C. G. Anderson's commentary in *Tacitus: Germania* (Bristol: Bristol Classical Press, 2001), 39–46, for an example of how such arguments played out in the 1930s (the original

is putting the Germani into a category of people (including, most notably, the Athenians) for whom the connection between environment and identity is particularly intimate. In doing so, he classes them with at least one group of people who may not be unequivocally admired by Romans, but are most definitely not considered barbarians (a term Tacitus avoids, as Gruen points out).[39] Here too, though, Tacitus is not endorsing the claim outright, since he ends the section with the remark, "which I intend neither to back up with arguments nor to refute: let each person either take or add credibility according to their own personality." He does not say, therefore, that the Germani *are* autochthonous. Rather, he says that the Germani proudly *claim* (*celebrant carminibus antiquis*) to have descended from an earthborn ancestor. In doing so, he puts into German mouths a sentiment that is familiar to those peoples who are already a part of the Pax Romana and that perhaps makes them seem less alien.

This is not the end of the Germani making claims that would be at home among their southern neighbors, though. Heracles was reported to have visited so many places that he must have needed a few lifetimes to do so, a trope that had become so common it was parodied by Lucian in the *True History* (1). When (again, in Tacitus's telling) the Germani claim to have been visited by Heracles and to still celebrate him when going into battle, Tacitus is giving them a place in the shared syncretic tradition of regions farther south. The reality of the situation—whether the Roman ethnographer was syncretizing some native hero-god figure—is unrecoverable. Second among those Greek mythological figures to have visited any and all locations is Odysseus; he too is mentioned by Tacitus. Not only did he visit, but—so Tacitus relates—he founded Asciburgium.[40] But Tacitus then undercuts all of this by distancing himself from the claims (see the second passage from *Germania* 2 above), thus reaffirming that it is not *he* who is making this claim, but the Germani themselves. To believe that the Germani came up with an origin story so very characteristic of Greek *poleis* strains credibility, but that does not mean that Tacitus is responsible for the initial imposition

---

date of publication is 1938). More recently, the question has been addressed by Norbert Wagner, "Lateinisch-Germanisch Mannus: Zu Tacitus, Germania C.2," *Historische Sprachforschung / Historical Linguistics* 107 (1994): 143–146; Dieter Timpe, *Romano—Germanica: Gesammelte Studien Zur Germania Des Tacitus* (Leipzig: Teubner, 1995), especially 1–61; Allan A. Lund, *Die Ersten Germanen: Ethnizität Und Ethnogenese* (Heidelberg: Winter, 1998), 58–85; H. Reichert, "Personennamen bei antiken Autoren als Zeugnisse für älteste westgermanische Endungen," *Zeitschrift für Deutsches Altertum und Deutsche Literatur* 132 (2003): 85–100; Greg Woolf, "Cruptorix and His Kind: Talking Ethnicity on the Middle Ground," in *Ethnic Constructs in Antiquity: The Role of Power and Tradition*, ed. Ton Derks and Nico Roymans (Amsterdam: Amsterdam University Press, 2009), 213–14.

39. Gruen, *Rethinking the Other*, 160–61.

40. A disputed toponym due to Greek letters to which one copyist seems to have been less than sanguine about including. See Anderson, *Tacitus: Germania*, apparatus criticus at *Germania* 3 and related note on page 50.

of Greek narrative patterns onto Germanic traditions. Whatever Tacitus says, the Germani were in close enough contact with their Southern neighbors that local legends could have been syncretized to fit the perceptions of Mediterranean visitors, especially if Tacitus's claims of Greek inscriptions in Asciburgium are true. Be that as it may, Tacitus here puts familiar sounding words in German mouths, and far from othering the Germani here, he includes them in the shared language of Greek and Roman ethnography and identity.

Tacitus's next chapter is the most infamous in the *Germania* for the way in which it was interpreted in later years as a positive and factual affirmation of German racial purity.

> Ipse eorum opinionibus accedo, qui Germaniae populos nullis aliis aliarum nationum conubiis infectos propriam et sinceram et tantum sui similem gentem exstitisse arbitrantur. Unde habitus quoque corporum, tamquam in tanto hominum numero, idem omnibus: truces et caerulei oculi, rutilae comae, magna corpora et tantum ad impetum valida: laboris atque operum non eadem patientia, minimeque sitim aestumque tolerare, frigora atque inediam caelo solove adsueverunt.

> I myself yield to the opinions of those who believe that the peoples of Germania, because they have been undone by no other marriages with other nations, have stood out as a special people, both without admixture and so very similar to itself. From that cause the appearance of their bodies, as much as can be the case in such a [large] number of people, is the same for all of them: piercing and sky-blue eyes, gingery hair, bodies that are big and only strong for spurts of effort: there is not the same endurance of labor or toil, and they have become least accustomed to tolerate thirst and heat, [but instead] cold and hunger due to their weather and soil (*Germania* 4).

The legacy of this passage overshadows what Tacitus's original intent may have been, and here environmental determinism can again aid in rethinking the way we read it.[41] Autochthony and purity combined with the subsequent description of German bodies provides a rationalization for what Tacitus later, in chapter 15, calls a "wondrous variability of character" (*mira diversitate naturae*)—Tacitus is referring here to the contradictory combination of bellicosity and laziness

---

41. Much ink has been spilled over that legacy, and interested readers would be best served by reference to Krebs, *Most Dangerous Book*, to which Martin A. Ruehl provides a lucid corrective rebuttal in "German Horror Stories: Teutomania and the Ghosts Of Tacitus," *Arion* 22 (2014): 129–90. To see what those assumptions from the 1930s looked like, one would be well served by consulting Anderson, *Tacitus: Germania*, 53–56.

attributed to Germani and, as Gruen points out, pretty much all inhabitants of the cold North.[42]

This passage, read against what came before it, presents purity as both strength and weakness—strength, because it implies that the positive qualities of the Germani are shared by all the Germani (for they are so *sui similes*), but weakness, because their faults are likewise shared uniformly. Tacitus moves swiftly from racial purity to a mixed bag of traits linked explicitly to climate: they are used to enduring cold and a lack of food because they live in a cold place lacking food, but they cannot endure its opposite (thirst and the sustained labor prized by farm-loving Romans) due to a combination of environment and lack of variation in their heritage. This is not simple environmental determinism but environmental determinism exacerbated by isolation. I think, far from the positive light in which the passage is usually read, this stands as a subtle endorsement of Roman conquest based not on political expedience (which, Gruen points out, Tacitus avoids) but rational analysis based on the natural sciences of the day.

Environmental determinism also offers something to the Roman anxiety about corrupting *luxuria*, a major concern voiced by Caesar, and even by Tacitus in the *Agricola* (See especially 6 and 15). Verbal echoes of Caesar keep that earlier ethnographer of the region ever in the reader's mind. Will the Romans destroy the very *virtus* they admire by breaking Germania's isolation? Not so in the *Germania*. Whereas in Caesar's *Gallic War*—and even in Tacitus's *Agricola*—*luxuria* was a formidable destroyer of northern *virtus*, here it is constructed as something to which the Germani have a natural resistance due, in part, to their land. These Germani are people grown out of the cold, unforgiving ground, uninterested in *luxuria* to the point that they treat precious metals with indifference.

> Terra etsi aliquanto specie differt, in universum tamen aut silvis horrida aut paludibus foeda ... ; satis ferax, frugiferarum arborum inpatiens, pecorum fecunda, sed plerumque improcera. Ne armentis quidem suus honor aut gloria frontis: numero gaudent, eaeque solae et gratissimae opes sunt. Argentum et aurum propitiine an irati di negaverint dubito. Nec tamen adfirmaverim nullam Germaniae venam argentum aurumve gignere: quis enim scrutatus est? Possessione et usu haud perinde adficiuntur. Est videre apud illos argentea vasa, legatis et principibus eorum muneri data, non in alia vilitate quam quae humo finguntur.

> Although the land differs a bit in its character, nevertheless the whole of it is either bristling with forests or grimy with swamps; sufficiently

---

42. See Gruen, *Rethinking the Other*, 159–61.

productive of grain, intolerant of fruit-bearing trees, productive for flocks, but for the most part [those flocks are] stunted. Not even the cattle herds have their own particular honor or glory of the brow: they rejoice in the number [of cattle], and that is their only and most valued wealth.[43] I hesitate to say whether generous or angry gods have denied them silver and gold. Nor can I even confirm that no part of Germany has given rise to a vein of silver or gold: for who has even looked?[44] For that reasons there is barely affixed any value to its possession or use. It is possible to see silver vases among them, given to their ambassadors and leaders as gifts, not held in any different value [lit. cheapness] as those which are sculpted from earth. (Tacitus, *Germania* 5)

This passage has often been read as a sneer at both the German landscape and German stupidity in not understanding the value of metals and money. But is it really so simple? Tacitus tells us later (*Germania* 26) that one of the virtues of German society is its lack of financial abuse.[45] Perhaps German disregard for the luxury of silver tableware is not a weakness, but a strength? If so, one ethical obstacle to expansion is removed, and I think again the logic of environment underlies the argument. If the German resistance to *luxuria* and corresponding commitment to *virtus* is engendered not by custom, but by the environment, then it will not be much altered by Roman conquest. Indeed, perhaps Romans can be inspired to imitate these valued qualities given exposure to a safe, pacified Germania.

It is no new thing to point out that Tacitus seems more concerned with using Germani to discuss Roman failings than with drawing a portrait of the Germani themselves. However, there is more to be made of the observation than simple caveats. Read against the political context of 98 AD and the imperialistic ambitions of Trajanic Rome, it is not only social commentary but rationalization for expansion. A particularly glaring example of Tacitus' "noble savage" Germani appears in chapter 20, where he says that Germani mothers do not use wetnurses and their children delay sexual activity and marriage, but such passages can be found throughout the treatise (see especially *Germania* 12, 14, 18–19, 26, 27). Roman moral guardians had all sorts of issues that they felt were eroding the

---

43. Comments about German cattle are not universally grim; Pliny, *Nat. hist.* 17.26.6, praises the pasturelands. But Tacitus, predictably, seems to be a partisan of Italian cattle.

44. Apparently, Pliny the Elder's sources did indeed look. *Nat. hist.* 34.2.5 discusses recently discovered German copper mines.

45. Faenus agitare et in usuras extendere ignotum; ideoque magis servatur quam si vetitum esset.

To scheme after profit and extend it into collecting interest is unknown [to the Germani]; for that reason, it is more absent than if it were forbidden.

mos maiorum and Roman society, by which they meant, of course, upper class Roman society. Sexual permissiveness, marital instability, the use of wet-nurses, limited family sizes, and predatory inheritance hunters all feature heavily among the complaints of Tacitus's contemporary elite Romans. If the Germani are naturally disposed to live free from these social ills, and if environment is a large factor in that *virtus*, then it follows that a solution to the perceived problems of contemporary Roman society would consist not only of enslaving and importing Germani into Rome but also of incorporating into the empire the very land that was so able to inculcate the *virtus* that upper class Romans found lacking in their contemporaries. It is not an overt argument; Tacitus is too subtle for that. But it is there, ready to be seized upon by hawkish readers who might very well come away thinking that it is their own idea.

Tacitus is not always (or often) so positive a critic of Germanic customs, and such passages also deserve some attention.

> Potui umor ex hordeo aut frumento, in quandam similitudinem vini corruptus: proximi ripae et vinum mercantur. Cibi simplices, agrestia poma, recens fera aut lac concretum: sine <u>apparatu</u>, sine <u>blandimentis</u> expellunt famem. Adversus sitim non eadem temperantia. Si indulseris ebrietati suggerendo quantum concupiscunt, haud minus facile vitiis quam armis vincentur.

> For drink, they have a liquid derived from barley or grain, spoiled into a certain semblance of wine: the riverbanks nearest to us purchase wine. Their foods are simple, rustic fruits, fresh wild game or cultured milk: without <u>fuss</u>, without <u>frills</u> they banish hunger. Against thirst there is not the same restraint. If you were to indulge them by enabling their drunkenness as much as they desire, they would be scarcely less easily conquered by vices as by arms (*Germania* 23).

Despite Tacitus's aversion to beer and smugness about the Germani's fondness for the beverage, the section is not wholly negative read against contemporary Roman handwringing about elaborate feasts and wasteful dining practices (*apparatus* and *blandimenta* are both buzzwords in this sort of discourse). Tacitus then transitions to the part of German dining he does *not* find admirable: binge drinking. Again, he sets his sights on an issue of concern also in Rome (excessive drinking), though one would not know it to read the dry observation that if you let a German drink as much as he is willing to do, one could just as easily conquer him with vice as with arms. It is a classic Tacitean line—terse, biting, and snide. But it is another thing important to this reading of the Germania: imperialistic. Tacitus walks a fine line here, praising the Germani as potential assets with the raw *virtus* and moral fiber unweakened by *luxuria*, but he must not go so far as to imply that Germania cannot be subdued. That is important, given the long history of Roman defeats along the Rhine and Danube, most glaring of which are the Teutoburg forest debacle and Domitian's still recent Dacian truce, but those come with a litany of other false inroads into this territory. The moments of German rashness, military simplicity, and poor provision serve two purposes: highlighting German strength while undercutting German invulnerability. Tacitus implies that the Germani—divided as they are by unruly brawling and quarrelsome, disorganized rule—however noble—seem to be in dire need of oversight. If, indeed, their bellicose nature is tied to the chilly damp of their homeland, their only hope is that those who hail from more temperate climes take charge.[46]

Whether Tacitus's *Germania* played any major role in spurring on the Dacian campaigns of Trajan is impossible to say; in many ways the conflict may have been inevitable. And environmental theory is certainly only one of many tropes Tacitus employs (or refuses to employ). Tan, for instance, in tracking Tacitus's avoidance of the standard structure of the geographical genre, sees an author much more interested in criticizing Domitian's Danubian failures than encouraging Nerva and Trajan's expansionist ambitions. But viewed against the tone and theoretical tendencies of its most likely source materials, Tacitus's choice of a hostile model that justifies a violent approach to Rome's neighbors to the North is clear. His audience would have been well able to hear these models in Tacitus's admittedly obtuse sketch of Germania's interior as well as his clear demarcation of the Roman borderlands. His description of the inhabitants gives his audience a vision of a people who are a tantalizing combination of warlike and simple; a contradictory portrait of the sort of people Rome both wished to conquer and wished to be. The environmental component of this regional character ensures that, even after conquest, Germania will retain its ability to foster *virtus* even under the pressure of *luxuria* flowing from the South.

Tacitus did not necessarily need to present such a fixed Germanic character to his audience. Strabo allows far greater room for cultural change independent of climactic undertow; his Helvetii can settle down to a life of contented dairy farming and his Celts can learn to live without the severed heads that had so horrified Poseidonios. Pliny the Elder, so far as we can tell, seems to have been similarly light in his use of environmental theory in service to anti-German

---

46. This way of thinking seems to have informed military policy to some extent: see Georgia Irby, "Climate and Courage," in *The Routledge Handbook of Identity and the Environment in the Classical and Medieval Worlds*, ed. Rebecca Futo Kennedy and Molly Ayn Jones-Lewis (London: Routledge, 2015), 247–65.

sentiment. It is Caesar, writing as a conqueror whose sources (especially Posei-donios and Eratosthenes) are used in service to an actively expansionist agenda, whom Tacitus invokes most deliberately in his sketched maps and ethnographic generalizations. Caesar, too, provides Tacitus with an environmental determin-ism drafted into service as a justification for war. For those educated Romans in his audience who were eager to make another attempt at German conquest, Tacitus's *Germania* would have seemed to provide justification for expanding Roman control past the Rhine and Danube. Considering the shape of the decades following Tacitus's publication of the *Germania*, more than a few members of the audience found confirmation for their existing biases toward the chilly North.

BIBLIOGRAPHY

Albertson, James. "Genesis 1 and the Babylonian Creation Myth." *Thought: Fordham University Quarterly* 37 (1962): 226–44.

Allan, Sarah. *The Shape of the Turtle.* Albany: State University of New York Press, 1991.

———. *The Way of Water and Sprouts of Virtue.* Albany: State University of New York Press, 2006.

Allen, Archibald. *The Fragments of Mimnermus.* Stuttgart: Steiner, 1993.

Allen, James P. "The Cosmology of the Pyramid Texts." Pages 1–28 in *Religion and Philosophy in Ancient Egypt.* Edited by James P. Allen. New Haven: Yale Egyptological Seminar, 1989.

Altenmüller, H. "Djed-Pfeiler." *LÄ* 1 (1975): 1100–1105.

Anderson, J. C. G. *Tacitus: Germania.* Bristol: Bristol Classical Press, 2001.

Arrighetti, Graziano. "Cosmologia Mitica di Omero e Esiodo." *SIFC* 15 (1966): 1–60.

Assmann, Jan. *Ägypten: Theologie und Frömmigkeit einer frühen Hochkultur.* Stuttgart: Kohlhammer, 1984.

———. *Der König als Sonnenpriester: Ein kosmographischer Begleittext zur kultischen Sonnenhymnik in thebanischen Tempeln und Gräbern.* Glückstadt: Augustin, 1970.

———. *Maât: L'Égypte pharaonique et l'idée de justice sociale.* Paris: Julliard, 1989.

———. "Schöpfung." *LÄ* 5 (1984): 677–90.

Atwell, James E. "An Egyptian Source for Genesis I." *JThS* 51 (2000): 441–77.

Aujac, Germaine. *Geminus, Introduction aux phénomènes.* Paris: Les Belles Lettres, 1975.

———. "Strabon et le stoïcisme." *Diotima* 11 (1983): 17–29.

Austin, Norman. *Archery at the Dark of the Moon.* Berkeley: University of California Press, 1975.

———. "The One and the Many in the Homeric Cosmos." *Arion* 1 (1973): 219–74.

Bakker, Frederik A. *Epicurean Meteorology: Sources, Method, Scope and Organization.* Leiden: Brill, 2016.

Ballabriga, Alain. *Les fictions d'Homère: L'invention mythologique et cosmographique dans L'Odyssée.* Paris: Presses universitaires de France, 1998.

———. *Le soleil et le Tartare: L'image mythique du monde en Grèce archaïque.* Paris: Éditions de l'École des hautes études en sciences sociales, 1986.

Barnes, Jonathan. "The Size of the Sun in Antiquity." *ACD* 25 (1989): 29–41.

Barrett, W. S. *Euripides Hippolytos*. Oxford: Oxford University Press, 1964.

Barton, Ian M. *Roman Domestic Buildings*. Exeter: University of Exeter Press, 1996.

Bauer, Brian S., Vania Smith-Oka, and Gabriel E. Cantarutti, trans. and eds. *Cristóbal de Molina: Account of the Fables and Rites of the Incas*. Austin: University of Texas Press, 2011.

Beagon, Mary. *Roman Nature: The Thought of Pliny the Elder*. Oxford: Clarendon, 1992.

Bentley, Richard. M. *Manilii Astronomicon*. London: Vaillant, 1739.

Beye, Charles Rowan. *Epic and Romance in the* Argonautica *of Apollonius*. Carbondale: Southern Illinois University Press, 1982.

Bianchetti, Serena. "Avieno, *Ora mar.* 80 ss.: Le colonne d'Eracle e il vento del nord." *Sileno* 16 (1990): 241–46.

Bing, Peter. "Aratus and His Audiences." *MD* 31 (1993): 99–109.

Bolton, James D. P. *Aristeas of Proconnesus*. Oxford: Clarendon, 1962.

Börtzler, Friedrich. "Zu den antiken Chaoskosmogonien." *Archiv für Religionswissenschaft* 28 (1930): 253–68.

Bowen, Alan C. "Papyrus Parisinus graecus 1." *EANS* 622.

———. *Simplicius on the Planets and Their Motions: In Defense of a Heresy*. Leiden: Brill, 2013.

Bowersock, Glen W. "The East-West Orientation of Mediterranean Studies and the Meaning of North and South in Antiquity." Pages 167–78 in *Rethinking the Mediterranean*. Edited by William V. Harris. Oxford: Oxford University Press, 2005.

Brumbaugh, Robert S. *The Philosophers of Greece*. Albany: State University of New York Press, 1964.

Brunner, Hellmut. "Die Grenzen von Zeit und Raum bei den Ägyptern." *AOF* 17 (1955): 141–45.

———. "Zum Raumbegriff der Ägypter." *Studium Generale: Zeitschrift für die Einheit der Wissenschaften* 10 (1957): 612–20.

Brunschvicg, Léon, ed. *Blaise Pascal: Pensées*. Paris: Vrin, 1904.

Bunbury, E. H. *History of Ancient Geography*. London: Murray, 1883.

Burkert, Walter. *Greek Religion*. Translated by John Raffan. Cambridge: Harvard University Press, 1985.

———. *The Orientalizing Revolution: Near Eastern Influence on Greek Culture in the Early Archaic Age*. Translated by Margaret E. Pinder. Cambridge: Harvard University Press, 1992.

Burnett, Anne Pippin. "Human Resistance and Divine Persuasion in Euripides' *Ion*." *CP* 57 (1962): 89–103.

Carlsen, Robert S., and Martin Prechtel. "The Flowering of the Dead: An Interpretation of Highland Maya Culture." *Man*, n.s., 26, no. 1 (1991): 23–42.

Casson, Lionel. *Ships and Seamanship in the Ancient World*. Princeton: Princeton University Press, 1971.

Cherniss, Harold. "The Characteristics and Effects of Presocratic Philosophy." *JHI* 12 (1951): 319–45.

Clark, Robert Thomas Rundle. *Myth and Symbol in Ancient Egypt*. London: Thames and Hudson, 1978.

Clarke, Katherine. *Between Geography and History: Hellenistic Constructions of the Roman World*. Oxford: Clarendon, 1999.

————. "In Search of the Author of Strabo's *Geography*." *JRS* 87 (1997): 92–110.

Clauss, James J. *The Best of the Argonauts: The Redefinition of the Epic Hero in Book One of Apollonius's* Argonautica. Berkeley: University of California Press, 1993.

Cohon, Robert. "Vergil and Pheidias: The Shield of Aeneas and of Athena Parthenos." *Vergilius* 37 (1991): 22–30.

Collins, John F. "Studies in Book One of the *Argonautica* of Apollonius Rhodius." PhD diss., Columbia University, 1967.

Conche, Marcel. *Anaximandre: Fragments et Témoignages*. Paris: Presses universitaires de France, 1991.

Conman, Joanne. "It's About Time: Ancient Egyptian Cosmology." *Studien zur Altägyptischen Kultur* 31 (2003): 33–71.

Cornelius, Sakkie. "Ancient Egypt and the Other." *Scriptura* 104 (2010): 322–40.

Cornell, Tim J. *The Fragments of the Roman Historians*. Oxford: Oxford University Press, 2013.

Couprie, Dirk L. *Heaven and Earth in Ancient Greek Cosmology*. Berlin: Springer, 2011.

Couprie, Dirk L., Robert Hahn, and Gérard Naddaf. *Anaximander in Context: New Studies in the Origins of Greek Philosophy*. Albany: State University of New York Press, 2003.

Coxon, A. H. *The Fragments of Parmenides*. Assen/Maastricht: Van Gorcum, 1986; Revised and expanded edition by Richard McKirahan and preface by Malcolm Scholfield. Las Vegas: Parmenides, 2009.

Cullen, Christopher. "A Chinese Eratosthenes of the Flat Earth: A Study of a Fragment of Cosmology in Huai Nan tzu 淮 南 子." *Bulletin of the School of Oriental and African Studies* 39, no. 1 (1976): 106–27.

Cusset, C. "Hegesianax." *EANS* 358.

Davies, Malcolm. *Poetarum melicorum graecorum fragmenta*. Oxford: Clarendon, 1991.

Dean, Dennis R. *James Hutton and the History of Geology*. Ithaca: Cornell University Press, 1992.

Demand, Nancy. "Epicharmus and Gorgias." *AJPh* 92 (1971): 453–63.

Dember, H. and M. Uibe. "Über die scheinbare Gestalt des Himmelsgewölbes." *Annalen der Physik* 360, no. 5 (1918): 387–96.

Derchain, Philippe. "Le rôle du roi d'Égypte dans le maintien de l'ordre cosmique." Pages 61–73 in *Le pouvoir et le sacré*. Edited by Luc de Heusch. Brussels: Université libre de Bruxelles, 1962.

De Santillana, Giorgio, and Hertha von Dechend. *Hamlet's Mill: An Essay on Myth and the Frame of Time*. Boston: Godine, 1969.

Dickey, Eleanor. *Ancient Greek Scholarship: A Guide to Finding, Reading, and Understanding Scholia, Commentaries, Lexica, and Grammatical Treatises, from Their Beginnings to the Byzantine Period*. Oxford: Oxford University Press, 2007.

Dicks, D. R. *Early Greek Astronomy to Aristotle*. Ithaca: Cornell University Press, 1970.

DiCosmo, Nicola. "The Northern Frontier in Pre-Imperial China." Pages 885–966 in *Cambridge History of Ancient China*. Edited by Michael Loewe and Edward L. Shaughnessy. New York: Cambridge University Press, 1999.

Diehl, Ernst, trans. *Procli Diadochi in Platonis Timaevm Commentaria*. Vol. 1. Leipzig: Teubner, 1903.

Diels, Hermann. *Simplicii in Aristotelis Physicorum Libros Quattuor Priores Commentaria*. Commentaria in Aristotelem Graeca 9. Berlin: Reimer, 1882.

———. *Simplicii in Aristotelis Physicorum Liber Quattuor Posteriores Commentaria*. Commentaria in Aristotelem Graeca 10. Berlin: Reimer, 1895.

Diels, Hermann, and Walter Kranz. *Die Fragmente der Vorsokratiker: Griechisch und deutsch*. 6th ed. Berlin: Weidmann, 1951.

Diggle, James, ed. *Euripides Phaethon; Edited with Prolegomena and Commentary*. Cambridge: Cambridge University Press, 1970.

———. *Studies on the Text of Euripides*: *Supplices, Electra, Heracles, Troades, Iphigenia in Tauris, Ion*. Oxford: Clarendon, 1981.

Dilke, O. A. W. *Greek and Roman Maps*. Ithaca: Cornell University Press, 1985.

Diller, Aubrey. "Agathemerus, *Sketch of Geography*." *GRBS* 16 (1975): 59–76.

———. *The Tradition of the Minor Greek Geographers*. Lancaster, PA: Lancaster Press, 1952.

Dittenberger, Wilhelm, and Karl Purgold. *Die Inschriften von Olympia*. Olympia 5. Berlin: Asher, 1896.

Dover, K. J., ed., *Aristophanes, Clouds*. Oxford: Clarendon Press, 1968.

Dueck, Daniela. "Pausanias of Damaskos." *EANS* 630–31.

———. *Strabo of Amasia: A Greek Man of Letters in Augustan Rome*. London: Routledge, 2000.

Dunand, Françoise, and Christiane Zivie-Coche, *Dieux et Hommes en Égypte*. Paris: Colin, 1991; repr. 2001.

———. *Gods and Men in Egypt*. Translated by David Lorton. Ithaca: Cornell University Press, 2004.

Dunbar, Nan, ed. *Aristophanes, Birds; Edited with Introduction and Commentary*. Oxford: Clarendon, 1995.

Dwyer, Eugene. "Augustus and the Capricorn." *MDAI(R)* 80 (1973): 59–67.

Eichholz, D. E. "The Shield of Aeneas: Some Elementary Notions." *PVS* 6 (1966–1967): 45–49.

Eliade, Mircea. *Cosmologie et Alchimie Babyloniennes*. Translated by Alain Paruit. Paris: Gallimard, 1991.

———. *Cosmologie şi alchimie babiloniană*. Bucharest: Vremea, 1937.

———. *Images and Symbols: Studies in Religious Symbolism*. Translated by Philip Mairet. London: Harvill, 1961.

———. *Images et Symboles*. Paris: Gallimard, 1952.

———. *Le Mythe de l'Éternel Retour: Archétypes et Répétition*. Paris: Gallimard, 1949.

———. *The Myth of the Eternal Return*. Translated by Willard R. Trask. New York: Pantheon, 1954.

———. "Psychologie et Histoire des Religions—A propos du Symbolisme du «Centre»." *Eranos Jahrbuch* 19 (1950 [1951]): 247–282. Reprinted with revisions as *Images et Symboles*. Paris: Gallimard, 1952. Translated by Philip Mairet as *Images and Symbols: Studies in Religious Symbolism*. New York: Sheed, Andrews, and McMeel, 1961 (repr. 1969).

Ellis, Robinson. "The Literary Relations of 'Longinus' and Manilius." *CR* 13 (1899): 294.

Endsjø, Dag Øistein. "Placing the Unplaceable: The Making of Apollonius' Argonautic Geography." *GRBS* 38 (1997): 373–85.

Enright, J. T. "The Eye, the Brain, and the Size of the Moon: Toward a Unified Oculo-motor Hypothesis for the Moon Illusion." Pages 59–122 in *The Moon Illusion*. Edited by Maurice Hershenson. Hillsdale, NJ: Erlbaum, 1989.

Euben, J. Peter. *Greek Tragedy and Political Theory*. Berkeley: University of California Press, 1986.

Euripides. *Hippolytos*. Edited by William S. Barrett. Oxford: Clarendon, 1964.

Evans, James, and J. Lennart Berggren. *Geminos's Introduction to the Phenomena: A Translation and Study of a Hellenistic Survey of Astronomy*. Princeton: Princeton University Press, 2006.

Evans, Rhiannon. "Ethnography's Freak Show." *Ramus* 28 (1999): 54–73.

Faber, Riemer Anne. "Vergil's Shield of Aeneas (*Aeneid* 8. 617–731) and the *Shield of Heracles*." *Mnemosyne* 53, no. 1 (2000): 48–57.

Faulkner, Raymond O. *A Concise Dictionary of Middle Egyptian*. Oxford: Oxford University Press, 1976.

Filehne, Wilhelm. "Die mathematische Ableitung der Form des scheinbaren Himmels-gewölbes." *Archiv für Physiologie: Physiologische Abteilung* 34, no. 1 (1912): 1–32.

Finkelberg, A. "On Cosmogony and Eypyrosis in Heraclitus." *AJPh* 119 (1998): 195–222.

Forrest, William G. G. "Colonization and the Rise of Delphi." *Historia* 6 (1957): 160–75.

Foster, Lynn V. *Handbook to Life in the Ancient Maya World*. Oxford: Oxford University Press, 2005.

Fracasso, Riccardo M. "Manifestazioni del simbolismo assiale nelle tradizione cinese antiche." *Numen* 28 (1981): 194–215.

Frankfort, Henri. *Kingship and the Gods*. 2nd ed. Chicago: University of Chicago Press, 1978.

Fraser, P. M. *Ptolemaic Alexandria*. 3 vols. Oxford: Clarendon, 1972.

Freidel, David, Linda Schele, and Joy Parker. *Maya Cosmos: Three Thousand Years on the Shaman's Path*. New York: Perennial, 1993.

Froidefond, Christian. *Le mirage égyptien dans la literature grecque d'Homère à Aristote* Paris: Ophrys, 1971.

Furley, David J. "Anaxagoras in Response to Parmenides." Pages 61–85 in *New Essays in Plato and the Pre-Socratics*. Canadian Journal of Philosophy Supplement 2. Edited by Roger A. Shiner and John King-Farlow. Guelph, Ontario: Canadian Assocation for Publishing in Philosophy, 1976.

———. "The Dynamics of the Earth: Anaximander, Plato, and the Centrifocal Theory." Pages 14–26 in *Cosmic Problems: Essays on Greek and Roman Philosophy of Nature*. Cambridge: Cambridge University Press, 1989.

———. *The Greek Cosmologists: The Formation of the Atomic Theory and Its Earliest Critics*. Vol. 1. Cambridge: Cambridge University Press, 1987.

Gallop, David. *Parmenides of Elea: Text and Translation with Introduction*. Toronto: University of Toronto Press, 1984.

Gardiner, Alan H. *Egyptian Grammar*. 3rd ed. London: Clarendon, 1957.

Gee, Emma. *Ovid, Aratus, and Augustus*. Cambridge: Cambridge University Press, 2000.

Gentili, Bruno, and Carolus Prato. *Poetarum elegiacorum testimonia et fragmenta*. 2nd ed. Leipzig: Teubner, 2002.

Gesenius, Wilhelm. *Gesenius's Hebrew and Chaldee Lexicon to the Old Testament Scriptures.* Translated by Simon P. Tregelles. New York: Wiley, 1893.

Geus, Klaus. "Der Widerstand gegen die Theorie von der Erde als Kugel: Paradigma einer Wissenschaftsfeindlichkeit in der heidnischen und christlichen Antike?" Pages 65–84 in *Exempla imitanda: Mit der Vergangenheit die Gegenwart bewältigen? Festschrift für Ernst Baltrusch zum 60. Geburtstag.* Edited by Monika Schuol, Christian Wendt, Julia Wilker, and Ernst Baltrusch. Göttingen: Vandenhoeck & Ruprecht, 2016.

Gibbs, Sharon L. *Greek and Roman Sundials.* New Haven: Yale University Press, 1976.

Gisinger, Friedrich. *Die Erdbeschreibung des Eudoxos von Knidos.* Leipzig: Teubner, 1921.

———. "Timosthenes von Rhodos [#3]." *RE,* second series, 6 (1937): 1310–22.

Goff, Barbara Elizabeth. "Euripides' *Ion* 1132–1165." *PCPhS* 34 (1988): 42–54.

González-Reimann, Luís. "Cosmic Cycles, Cosmology, and Cosmography." Pages 411–28 in vol. 2 of *Brill's Encyclopedia of Hinduism.* Edited by Knut A. Jacobsen. Leiden: Brill, 2009.

Graham, Daniel. *The Texts of Early Greek Philosophy: The Complete Fragments and Selected Testimonies of the Major Presocratics.* Cambridge: Cambridge University Press, 2010.

Green, Peter, trans. *Apollonios of Rhodes: The Argonautika.* Berkeley: University of California Press, 2007.

Grégoire, Henri, ed. *Euripide: Tome III.* Paris: Les Belles Lettres, 1950.

Grenfell, Bernard P., and Arthur S. Hunt. *Greek Papyri, Series II: Classical Fragments and Other Greek and Latin Papyri.* Oxford: Clarendon, 1897.

Gruen, Erich S. *Rethinking the Other in Antiquity.* Princeton: Princeton University Press, 2012.

Grüninger, Gerhart. "Untersuchungen zur Persönlichkeit des älteren Plinius: Die Bedeutung wissenschaftlicher Arbeit in seinem Denken." PhD diss., University of Freiburg, 1976.

Gurval, Robert Alan. *Actium and Augustus: The Politics and Emotions of Civil War.* Ann Arbor: University of Michigan Press, 1998.

Guthrie, William K. C. *A History of Greek Philosophy 1: The Earlier Presocratics and the Pythagoreans.* Cambridge: Cambridge Univerity Press, 1962.

———. *A History of Greek Philosophy 2: The Presocratic Tradition From Parmenides to Democritus.* Cambridge: Cambridge University Press, 1965.

Hahn, Robert. *Anaximander and the Architects: The Contributions of Egyptian and Greek Architectural Technologies to the Origins of Greek Philosophy.* Albany: State University of New York Press, 2001.

Hallpike, Christopher R. *The Foundations of Primitive Thought.* Oxford: Clarendon, 1979.

Halpern, Baruch. "The Assyrian Astronomy of Genesis 1 and the Birth of Milesian Philosophy." *Eretz-Israel* 27 (2003): 74–83.

Hardie, Philip R. "Imago Mundi: Cosmological and Ideological Aspects of the Shield of Achilles." *JHS* 105 (1985): 11–31.

———. *Vergil's Aeneid: Cosmos and Imperium.* Oxford: Clarendon, 1986.

Harding, Phillip. *Didymos on Demosthenes.* Oxford: Clarendon, 2006.

Harley, John B., and David A. Woodward. *Cartography in Prehistoric, Ancient, and Medieval Europe and the Mediterranean.* Chicago: University of Chicago Press, 1987.

Harrison, Stephen J. "The Survival and Supremacy of Rome: The Unity of the Shield of Aeneas." *JRS* 87 (1997): 70–76.

Hartog, François. *The Mirror of Herodotus.* Translated by Janet Lloyd. Berkeley: University of California Press, 1988.

Hawke, Jason. "Number and Numeracy in Early Greek Literature." *SyllClass* 19 (2008): 1–76.

Hayduck, Michael. *Alexandri in Aristotelis Meteorologicorum Libros Commentaria.* Berlin: Reimer, 1899.

Heather, Peter J. *Empires and Barbarians: The Fall of Rome and the Birth of Europe.* New York: Oxford University Press, 2012.

Heiberg, Johan L. *Simplicii in Aristotelis de Caelo Commentaria.* Berlin: Reimer, 1894.

Heidel, William A. "Anaximander's Book: The Earliest Known Geographical Treatise." *Proceedings of the American Academy of the Arts and Sciences* 56 (1921): 239–88.

———. *The Frame of the Ancient Greek Maps.* New York: American Geographical Society, 1937.

Heimpel, Wolfgang. "The Sun At Night and the Doors of Heaven in Babylonian Texts." *Journal of Cuneiform Studies* 38 (1986): 127–51.

Helck, Wolfgang, and Eberhard Otto, eds. *Lexikon der Ägyptologie.* 7 vols. Wiesbaden: Harrassowitz, 1975–1990.

Hinckley, Lois V., and Michelle Thorne. "The Shields of Achilles and Aeneas in Dialogue." *NECJ* 21 (1993–1994): 149–55.

Hippocrates. *Airs, Eaux, Lieux.* Edited by Jacques Jouanna. Série grecque 374. Paris: Les Belles Lettres, 1996.

Hölbl, Günther. *A History of the Ptolemaic Empire.* Translated by Tina Saavedra. London: Routledge, 2001.

Hölscher, Tonio. "Ein römischer Stirnziegel mit Victoria und Capricorn." *JRGZ* 12 (1965): 59–73.

Hölscher, Uvo. "Anaximander und die Anfänge der Philosophie (II)." *Hermes* 81 (1953): 385–418.

Holder, Paul. "Auxiliary Deployment in the Reign of Hadrian." *BICS* 46 (2003): 101–45.

Hopkins, E. Washburn. "Mythological Aspects of Trees and Mountains in the Great Epic." *JAOS* 30 (1910): 347–74.

Horden, Peregrine, and Nicolas Purcell. *The Corrupting Sea: A Study of Mediterranean History.* Oxford: Blackwell, 2000.

Hornung, Erik. "Chaotische Bereiche in der geordneten Welt." *ZÄS* 81 (1956): 28–32.

Horowitz, Wayne. "The Babylonian Map of the World." *Iraq* 50 (1988): 147–65.

———. *Mesopotamian Cosmic Geography.* Winona Lake, IN: Eisenbrauns, 1998.

Hubbard, Thomas K. *The Pindaric Mind: A Study of Logical Structure in Early Greek Poetry.* Leiden: Brill, 1985.

Huffman, Carl A. *Philolaus of Croton.* Cambridge: Cambridge University Press, 1993.

Hurst, André. "Géographes et poètes: Le cas d'Apollonios de Rhodes." Pages 279–88 in *Sciences exactes et sciences appliquées à Alexandrie.* Edited by Gilbert Argoud and Jean-Yves Guillaumin. Saint-Étienne: Publications de l'Université de Saint-Étienne, 1998.

Hurwit, Jeffrey M. *The Art and Culture of Early Greece: 1100–480 B.C.* Ithaca: Cornell University Press, 1985.

Imhof, Max. *Euripides' Ion: Eine literarische Studie*. Bern: Francke, 1966.

Inwood, Brad. *The Poem of Empedocles*. 2nd ed. Toronto: University of Toronto Press, 2001.

Irby, Georgia. "Climate and Courage." Pages 247–65 in *The Routledge Handbook of Identity and the Environment in the Classical and Medieval Worlds*. Edited by Rebecca Futo Kennedy and Molly Ayn Jones-Lewis. London: Routledge, 2015.

Irby-Massie, Georgia. "M. Vipsanius Agrippa." *EANS* 830.

Isaac, Benjamin. *The Invention of Racism in Classical Antiquity*. Princeton: Princeton University Press, 2013.

Janowski, Bernd. "Vom näturlich zum symbolisch Raum: Aspekte der Raumwahrneh-mung in Alten Testament." Pages 51–64 in *Wahrnehmung und Erfassung geographi-scher Räume in der Antike*. Edited by Michael Rathmann, Mainz: von Zabern, 2007.

Jones, Alexander. "The Stoics and the Astronomical Sciences." Pages 342–44 in *The Cambridge Companion to the Stoics*. Edited by Brad Inwood. Cambridge: Cambridge University Press, 2003.

Jones, William H. S. *Hippocrates*. Vol. 2. Loeb Classical Library 148. Cambridge: Harvard University Press, 1923.

Jones-Lewis, Molly Ayn. "Poison: Nature's Argument for the Roman Empire in Pliny the Elder's *Naturalis Historia*." *CW* 106 (2012): 51–74.

———. "Tribal Identity in the Roman World: The Case of the Psylloi." Pages 192–209 in *The Routledge Handbook of Identity and the Environment in the Classical and Medieval Worlds*. Edited by Rebecca Futo Kennedy and Molly Jones-Lewis. London: Routledge, 2015.

Jouanna, Jacques. "Water, Health, and Disease in the Hippocratic Treatise *Airs, Waters, Places*." Pages 155–72 in *Greek Medicine from Hippocrates to Galen*. Edited by Philip van der Eijk. Translated by Neil Alles. Leiden: Brill, 2012.

Kahanov, Yaacov. "Ma'agan-Michael ship (Israel)." Pages 155–60 in *Construction navale maritime et fluviale: Approches archéologique, historique et ethnologique*. Edited by Patrice Pomey and Eric Rieth. Archaeonautica 14. Paris: CNRS, 1999.

Kahn, Charles H. *Anaximander and the Origins of Greek Cosmology*. New York: Columbia University Press, 1960.

Kannicht, Richard. *Tragicorum Graecorum Fragmenta*. Vol. 5. Göttingen: Vandenhoeck & Ruprecht, 2004.

Kaplan, Philip. "Hekataios of Abdera." *EANS* 361.

———. "Ktesias of Knidos." *EANS* 496.

———. "Skulax of Karuanda." *EANS* 745–46.

———. "Skulax of Karuanda, pseudo." *EANS* 746.

Karenga, Maulana. *Maat, The Moral Ideal in Ancient Egypt: A Study in Classical African Ethics*. New York: Routledge, 2004.

Kartunnen, Klaus. "ΚΥΝΟΚΕΦΑΛΟΙ and ΚΥΝΑΜΟΛΓΟΙ in Classical Ethnography." *Arctos* 18 (1984): 31–36.

Kaufman, Lloyd, and James H. Kaufman. "Explaining the Moon Illusion." *Proceedings of the National Academy of Sciences of the United States of America* 97 (2000): 500–505.

Kaufman, Lloyd, and Irvin Rock. "The Moon Illusion, I." *Science* 136 (1962): 953–61.

Kaufman, Lloyd, V. Vassiliades, R. Noble, R. Alexander, J. Kaufman, S. Edlund. "Perceptual Distance and the Moon Illusion." *Spatial Vision* 20 (2007): 155–75.

Keay, Simon J. *Roman Spain*. Berkeley: University of California, 1988.

Keightley, David N. *The Ancestral Landscape: Time, Space, and Community in Late Shang China, ca. 1200–1045 B.C.* Berkeley: University of California Press, 2000.

Kennedy, Rebecca Futo. "Airs, Waters, Metals, Earth: People and Land in Archaic and Classical Greek Thought." Pages 9–28 in *The Routledge Handbook of Identity and the Environment in the Classical and Medieval Worlds*. Edited by Rebecca Futo Kennedy and Molly Jones-Lewis. London: Routledge, 2015.

Keppie, Lawrence. *The Making of the Roman Army: From Republic to Empire*. Norman: University of Oklahoma Press, 1998.

Keyser, Paul T. "Baiton." *EANS* 186.

———. "Diognetos." *EANS* 254.

———. "From Myth to Map: The Blessed Isles in the First Century BC." *AncW* 24 (1993): 149–68.

———. "The Geographical Work of Dikaiarchos." Pages 353–72 in *Dicaearchus of Messana: Text, Translation, and Discussion*. Edited by W. W. Fortenbaugh and Eckahrt Schütrumpf. New Brunswick, NJ: Transaction Publishers, 2001.

———. "Greek Geography of the Western Barbarians." Pages 37–70 in *The Barbarians of Ancient Europe: Realities and Interactions*. Edited by Larissa Bonfante. Cambridge: Cambridge University Press, 2011.

———. "Kleon of Surakousai." *EANS* 481.

———. "The Name and Nature of Science: Authorship in Social and Evolutionary Context." Pages 17–61 in *Writing Science: Medical and Mathematical Authorship in Ancient Greece*. Edited by Markus Asper. Berlin: de Gruyter, 2013.

———. "Philonides of Khersonesos." *EANS* 659.

———. "Xenophanes' Sun on Trojan Ida." *Mnemosyne* 45 (1992): 299–311.

Keyser, Paul T., and Georgia L. Irby-Massie. "Diodoros of Tarsos." *EANS* 249–50.

———, eds. *Encyclopedia of Ancient Natural Scientists: The Greek Tradition and Its Many Heirs*. London: Routledge, 2008.

Kidd, Douglas A. *Aratus, Phaenomena: Edited with Introduction, Translation and Commentary*. Cambridge: Cambridge University Press, 1997.

Kidd, Ian G. *Posidonius, Vol II: The Commentary*. Cambridge: Cambridge University Press, 1988.

Kirk, G. S., J. E. Raven, and M. Schofield. *The Presocratic Philosophers: A Critical History with a Selection of Texts*. 2nd ed. Cambridge: Cambridge University Press, 1983.

Kister, Menahem. "*Tohu wa-Bohu*, Primordial Elements and *Creatio ex Nihilo*." *Jewish Studies Quarterly* 14 (2007): 229–56.

Klimkeit, Hans J. "Spatial Orientation in Mythical Thinking as Exemplified in Ancient Egypt: Considerations Toward a Geography of Religions." *HR* 14 (1975): 266–81.

Koniaris, George Leonidas, ed. *Maximus Tyrius: Philosophumena—ΔΙΑΛΕΞΕΙΣ*. Berlin: de Gruter, 1995.

Korenjak, Martin. "*Italiam contra Tiberinaque longe/Ostia*: Virgil's Carthago and Eratosthenian Geography." *CQ* 54 (2004): 646–49.

Kraft, K. "Zum Capricorn auf den Münzen des Augustus." *JNG* 17 (1967): 17–27.

Krebs, Christopher B. *A Most Dangerous Book: Tacitus's* Germania *From the Roman Empire to the Third Reich*. New York: Norton, 2012.

Kuelzer, Andreas. "Byzantine Geography." Pages 921–42 in *Oxford Handbook of Science and Medicine in the Classical World*. Edited by Paul T. Keyser with John Scarborough. New York: Oxford University Press, 2018.

Kurth, Dieter. "Manu." *LÄ* 3 (1980): 1185–86.

———. "Nut." *LÄ* 4 (1982): 533–41.

Laskaris, J. "Hippokratic Corpus, Airs, Waters, Places." *EANS* 406.

Lasserre, François. *Die Fragmente des Eudoxos von Knidos*. Berlin: de Gruyter, 1966.

Laurent, Jérôme. "Strabon et la philosophie stoïcienne." *ArchPhilos* 71 (2008): 111–27.

Lawall, Gilbert. "Apollonius' *Argonautica*: Jason as Anti-Hero." *YCS* 19 (1966): 121–69.

Lehoux, Daryn. "Poseidonios of Apameia (ca. 110–ca. 51 BCE)." *EANS* 691–92.

Leimbach, R. "Euripides *Ion*: Eine Interpretation." PhD diss., Frankfurt am Main, 1971.

Lesher, J. H. *Xenophanes of Colophon: Fragments; A Text and Translation with a Commentary*. Toronto: University of Toronto Press, 1992.

Lesko, Leonard H. "Ancient Egyptian Cosmogonies and Cosmologies." Pages 88–122 in *Religion in Ancient Egypt: Gods, Myths, and Personal Practice*. Edited by Byron E. Shafer. Ithaca: Cornell University Press, 1991.

Levin, Donald N. "*Diplax Porphuree*." *RFIC* 93 (1970): 17–36.

Lewis, C. S. *The Discarded Image: An Introduction to Medieval and Renaissance Literature*. Cambridge: Cambridge University Press, 1964.

Lewis, Mark Edward. *The Construction of Space in Early China*. Albany: State University of New York Press, 2006.

———. *The Flood Myths of Early China*. Albany: State University of New York Press, 2006.

Littré, Émile. *Oeuvres complètes d'Hippocrate: Traduction nouvelle avec le text grec en regard*. Paris: Bailliere, 1839–1861.

Lloyd, G. E. R. *The Ambitions of Curiosity*. Cambridge: Cambridge University Press, 2002.

———. *Greek Science After Aristotle*. New York: Norton, 1973.

———. *Magic, Reason, and Experience*. Cambridge: Cambridge University Press, 1979.

———. *Polarity and Analogy*. Cambridge: Cambridge University Press, 1966.

Lonie, Iain M., ed. *The Hippocratic Treatises "On Generation," "On the Nature of the Child," "Diseases 4": A Commentary*. Berlin: de Gruyter, 1981.

Loprieno, Antonio. *Topos und Mimesis: Zum Ausländer in der ägyptischen Literatur* Wiesbaden: Harrassowitz, 1988.

Lozovsky, N. "Kosmas of Alexandria, Indikopleustes." *EANS* 487.

Luckiesh, M. "The Apparent Form of the Sky-Vault." *Journal of the Franklin Institute* 191 (1921): 259–63.

Lund, Allan A. *Die Ersten Germanen: Ethnizität Und Ethnogenese*. Heidelberg: Winter, 1998.

Lynch, David K. "Optics of Sunbeams." *Journal of the Optical Society of America* A 4.3 (1987): 609–11.

Lynn, Chris, and Dean A. Miller. "The Shield Beyond All Words to Describe: Trifunctional Patterns Located on the Shield of Aeneas?" *JIES* 31 (2003): 391–419.

MacCormack, Geoffrey. "Natural Law and Cosmic Harmony in Traditional Chinese Thought." *Ratio Juris* 2 (1989): 254–73.

Mair, Victor H. *Wandering on the Way: Early Taoist Tales and Parables of Chuang Tzu*. New York: Bantam, 1994.

Major, John S. *Heaven and Earth in Early Han Thought: Chapters Three, Four, and Five of the Huainanzi*. Albany: State University of New York Press, 1993.

———. "Myth, Cosmology, and the Origins of Chinese Science." *Journal of Chinese Philosophy* 5 (1978): 1–20.

Marcotte, Didier. *Géographes Grecs*. Vol. 1, *Introduction générale; Ps.-Scymnos; Circuit de la Terre*. Paris: Les Belles Lettres, 2000.

Marcovich, Miroslav. *Heraclitus*. Mérida, Venezuela: Los Andes University Press, 1967.

Margarida, Ana Arruda. "Phoenician Colonization on the Atlantic Coast of the Iberian Peninsula." Pages 113–30 in *Colonial Encounters in Ancient Iberia*. Edited by Michael Dietler and Carolina López-Ruiz. Chicago: University of Chicago Press, 2009.

Martin, Karl. "Urhügel." *LÄ* 6 (1986): 873–75.

Mastronarde, Donald John. "Iconography and Imagery in Euripides' *Ion*." *CSCA* 8 (1975): 163–76.

Mayor, Adrienne. *The First Fossil Hunters: Dinosaurs, Mammoths, and Myth in Greek and Roman Times*. Princeton: Princeton University Press, 2011.

———. *The First Fossil Hunters: Paleontology in Greek and Roman Times*. Princeton: Princeton University Press, 2000.

McKeown, Jennifer. "The Symbolism of the Djed-pillar in *The Tale of King Khufu and the Magicians*." *Trabajos de Egiptología / Papers on Ancient Egypt* 1 (2002): 55–68.

Mendell, H. "Meton of Athens." *EANS* 551–52.

Merkelbach, Reinhold, and M. L. West. *Hesiodi Theogonia, Opera det Dies, Scutum, Fragmenta Selecta*. 3rd ed. Oxford: Clarendon, 1990.

Merker, Irwin L. "The Ptolemaic Officials and the League of the Islanders." *Historia* 19 (1970): 141–60.

Merriam, Carol Una. "An Examination of Jason's Cloak (Apollonius Rhodius, *Argonautica* 1, 730–68)." *Scholia* 2 (1993): 69–80.

Meyer, J. "Aëtios." *EANS* 37–38.

Miller, Albert, and Hans Neuberger. "Investigations into the Apparent Shape of the Sky." *Bulletin of the American Meteorological Society* 26 (1945): 212–16.

Mittenhuber, Florian. "Die Naturphänomene des hohen Nordens in den kleinen Schriften des Tacitus." *MH* 60 (2003): 44–59.

Moraux, Paul. "Anecdota Graeca Minora II: Über die Winde." *ZPE* 41 (1981): 43–58.

Morenz, Siegfried. *Egyptian Religion*. Translated by Ann E. Keep. Ithaca: Cornell University Press, 1973.

Morrison, J. S. "The Shape of the Earth in Plato's *Phaedo*." *Phronesis* 4 (1959): 101–19.

Muhs, Brian P. *The Ancient Egyptian Economy, 3000–30 BCE*. Cambridge: Cambridge University Press, 2016.

Müller, Karl. *Geographi Graeci Minores*. 2 vols. Paris: Didot, 1855–1882.

Müller, Klaus E. *Geschichte der Antiken Ethnographie und Ethnologischen Theoriebildung*. 2 vols. Wiesbaden: Steiner, 1972, 1980.

Murray, William M., and Photios M. Petsas. *Octavian's Campsite Memorial for the Actian War*. Philadelphia: American Philosophical Society, 1989.

Myres, John L. *Herodotus: Father of History*. Oxford: Clarendon, 1953.

Nakayama, Shigeru. *A History of Japanese Astronomy: Chinese Background and Western Impact*. Cambridge: Harvard University Press, 1969 ..

Needham, Joseph. *Science and Civilization in China*. Cambridge: Cambridge University Press, 1959.

Needham, Joseph, and Colin A. Ronan. "Chinese Cosmology." Pages 25–35 in *Encyclopedia of Cosmology*. Edited by Noriss S. Hetherington. New York: Garland, 1993.

Neugebauer, Otto. *Astronomy and History: Selected Essays*. New York: Springer, 1983.

Neugebauer, Otto, and R. A. Parker. *Egyptian Astronomical Texts I: The Early Decans*. London: Lund Humphries, 1960.

Newsome, Elizabeth. *Trees of Paradise and Pillars of the World: The Serial Stelae Cycle of '18-Rabbit-God K,' King of Copan*. Austin: University of Texas Press, 2001.

Nichols, Andrew. *Ctesias, On India: Introduction, Translation, and Commentary*. London: Bloomsbury, 2011.

Nicolet, Claude. *Space, Geography, and Politics in the Early Roman Empire*. Ann Arbor: University of Michigan Press, 1991.

O'Gorman, Ellen. "No Place Like Rome: Identity and Difference in the *Germania* of Tacitus." Pages 95–118 in *Tacitus*. Edited by Rhiannon Ash. London: Bloomsbury, 2006.

Osborne, Catherine. *Rethinking Early Greek Philosophy: Hippolytos of Rome and the Presocratics*. London: Duckworth, 1970.

Östenberg, Ida. "Demonstrating the Conquest of the World: The Procession of Peoples and Rivers on the Shield of Aeneas and the Triple Triumph of Octavian in 29 B.C." *Orom* 24 (1999): 155–62.

Otto, Eberhard. "Bachu." *LÄ* I (1975): 574.

Ottone, Gabriella. "Strabone e la critica a Timostene di Rodi: Un frammento di Polibio (XII.1.5) testimone del Περὶ λιμένων?" *Syngraphé* 4 (2002): 153–71.

Page, D. L. *Poetae melici Graeci*. Oxford: Clarendon, 1962.

Pankenier, David W. *Astrology and Cosmology in Early China: Conforming Earth to Heaven*. Cambridge: Cambridge University Press, 2013.

———. "The Cosmo-Political Background of Heaven's Mandate." *Early China* 20 (1995): 121–76.

———. "Heaven-Sent: Understanding Cosmic Disaster in Chinese Myth and History." Pages 187–97 in *Natural Catastrophes During Bronze Age Civilisations: Archaeological, Geological, Astronomical, and Cultural Perspectives*. Edited by Benny J. Peiser, Trevor Palmer, and Mark E. Bailey. Oxford: Archaeopress, 1998.

Pascal, Blaise. *Pensées*. Pages 541–1688 in *Oeuvres complètes*. Edited by Michel Le Guern. Paris: Gallimard, 2000.

Pendrick, Gerard J. "Antiphon of Athens." *EANS* 99.

———. *Antiphon the Sophist*. Cambridge: Cambridge University Press, 2002.

Picot, Jean-Claude. "L'Image du ΠΝΙΓΕΥΣ dans les *Nuées*: Un Empédocle au charbon." Pages 113–29 in *Comédie et Philosophie: Socrate et les « Présocratiques » dans les Nuées d'Aristophane*. Edited by André Laks and Rosella Saetta Cottone. Paris: Presses de l'École normale supérieure, 2013.

Plofker, Kim. "Humans, Demons, Gods and Their Worlds: The Sacred and Scientific Cosmologies of India." Pages 32–42 in *Geography and Ethnography: Perceptions of the World in Pre-Modern Societies*. Edited by Kurt A. Raaflaub and Richard J. A. Talbert. West Sussex: Wiley, 2013.

Podlecki, A. J. *The Life of Themistocles: A Critical Survey of the Literary and Archaeological Evidence*. Montreal: McGill-Queen's University Press, 1975.

Pomponius Mela. *Chorographie*. Translated by Alain Silberman. Paris: Les Belles Lettres, 1988.

———. *De chorographia libri tres*. Translated and edited by Piergiorgio Parroni. Rome: Edizioni di storia e letteratura, 1984.

———. *Kreuzfahrt durch die Alte Welt*. Translated and edited by Kai Brodersen. Darmstadt: Wissenschaftliche Buchgesellschaft, 1994.

Pongratz-Leisten, Beate. "The Other and the Enemy in the Mesopotamian Conception of the World." Pages 195–231 in *Mythology and Mythologies: Methodological Approaches to Intercultural Influences*. Edited by Robert Whiting. Helsinki: Neo-Assyrian Text Corpus Project, 2001.

Putnam, Michael C. J. *Vergil's Epic Designs: Ecphrasis in the Aeneid*. New Haven: Yale University Press, 1998.

Raaflaub, Kurt A., and Richard J. A. Talbert. *Geography and Ethnography: Perceptions of the World in Pre-Modern Societies*. New York: Wiley, 2012.

Radt, Stefan. *Strabons Geographika*. 10 vols. Göttingen: Vandenhoeck & Ruprecht, 2002–2011.

———. *Tragicorum graecorum fragmenta*. 2nd ed. Göttingen: Vandenhoeck & Ruprecht 1999.

Rao, D. Venkateswara. "Effect of Illumination on the Apparent Shape of the Sky." *Proceedings of the Indian Academy of Sciences, Section A* 25, no. 1 (1947): 34–42.

———. "Variation of the Apparent Shape of the Sky with Intensity of Illumination." *Current Science* 15, no. 2 (1946): 40–41.

Raphals, Lisa. "A 'Chinese Eratosthenes' Reconsidered: Chinese and Greek Calculations and Categories." *East Asian Science, Technology, and Medicine* 19 (2002): 10–60.

Rappenglueck, Michael. "A Palaeolithic Planetarium Underground: The Cave of Lascaux." *Migration and Diffusion* 5 (2004): 93–119.

Rathmann, Michael. *Wahrnehmung und Erfassung geographischer Räume in der Antike*. Mainz: von Zabern, 2007.

Read, Kay A. "Sacred Commoners: The Motion of Cosmic Powers in Mexican Rulership." *HR* 34 (1994): 39–69.

Reichert, H. "Personennamen bei antiken Autoren als Zeugnisse für älteste westgermanische Endungen." *Zeitschrift für Deutsches Altentum und Deutsches Literatur* 132 (2003): 85–100.

Reimann, Eugen. "Die scheinbare Vergrößerung der Sonne und des Mondes am Horizont." *Zeitschrift für Psychologie* 30 (1902): 1–38.

Riesenberg, Saul H. "The Organization of Navigational Knowledge on Puluwat." Pages 91–128 in *Pacific Navigation and Voyaging*. Edited by Ben R. Finney. Wellington, New Zealand: Polynesian Society, 1976.

Rihll, Tracey Elizabeth. *Greek Science*. Oxford: Oxford University Press, 1999.

Robinson, Arthur H. "The Uniqueness of the Map." *American Cartographer* 5 (1978): 5–7.

Robinson, T. M. *Heraclitus: Fragments*. Toronto: University of Toronto Press, 1987.

Rochberg, Francesca. "The Expression of Terrestrial and Celestial Order in Ancient Mesopotamia." Pages 9–46 in *Ancient Perspectives: Maps and Their Place in Mesopotamia, Egypt, Greece and Rome*. Edited by Richard J. A. Talbert. Chicago, University of Chicago Press, 2012.

———. "A Short History of the Waters of the Firmament." Pages 227–44 in *From the Banks of the Euphrates: Studies in Honor of Alice Louise Slotsky*. Edited by Micah Ross. Winona Lake, IN: Eisenbrauns, 2008.

Rochberg-Halton, Francesca. "Mesopotamian Cosmology." Pages 398–407 in *Encyclopedia of Cosmology*. Edited by Noriss S. Hetherington. New York: Garland, 1993.

Rock, Irvin, and Lloyd Kaufman. "The Moon Illusion, II." *Science* 136 (1962): 1023–31.

Rodríguez-Noriega Guillién, Lucía. *Epicarmo de Siracusa: Testimonios y Fragmentos*. Oviedo: Universidad de Oviedo, Servicio de Publicaciones, 1996.

———. "Epikharmos of Surakousai." *EANS* 291–92.

Roller, Duane W. *Ancient Geography: The Discovery of the World in Ancient Greece and Rome*. London: Tauris, 2015.

———. *Eratosthenes' Geography*. Princeton: Princeton University Press, 2010.

———. *The Geography of Strabo: An English Translation with Introduction and Notes*. Cambridge: Cambridge University Press, 2014.

———. "Phoenician Exploration." Pages 645–53 in *Oxford Handbook of the Phoenician and Punic Mediterranean*. Edited by Carolina López-Ruiz and Brian Doak. Oxford: Oxford University Press, 2019.

———. "Seleukos of Seleukeia." *Antiquite Classique* 74 (2005): 111–18.

———. *Through the Pillars of Herakles: Greco-Roman Exploration of the Atlantic*. London: Routledge, 2006.

———. *The World of Juba II and Kleopatra Selene: Royal Scholarship on Rome's African Frontier*. London: Routledge, 2003.

Romer, Frank E. *Pomponius Mela's Description of the World*. Ann Arbor: University of Michigan Press, 1998.

Romero, Aldemaro. "It's a Wonderful Hypogean Life: A Guide to the Troglomorphic Fishes of the World.: Pages 13–41 in *The Biology of Hypogean Fishes*. Edited by Aldemaro Romero. Dordrecht: Kluwer, 2001.

Romm, James S. "Dragons and Gold at the Ends of the Earth: A Folktale Motif Developed by Herodotus." *Merveilles & Contes* 1 (1987): 45–54.

———. *The Edges of the Earth in Roman Thought*. Princeton: Princeton University Press, 1992.

———. "Herodotus and Mythic Geography: The Case of the Hyperboreans." *TAPhA* 119 (1989): 97–113.

Rood, Tim. "Mapping Spatial and Temporal Distance in Herodotus and Thucydides." Pages 101–20 in *New Worlds from Old Texts: Revisting Ancient Space and Place*. Edited by Elton Barker, Stefan Bouzarovski, C. B. R. Pelling, and Leif Isaksen. Oxford: Oxford University Press, 2016.

Roscher, Wilhelm H. *Der Omphalosgedanke bei verschiedenen Völkern, besonders den semitischen: Ein Beitrag zur vergleichenden Religionswissenschaft, Volkskunde und Archäologie = Berichte über die Verhandlungen der Sächsischen Gesellschaft der Wissenschaften zu Leipzig: Philologisch-historische Klasse* 70. Hildesheim: Olms, 1974.

Ross, Helen E. "Cleomedes (c. 1st century AD) on the Celestial Illusion, Atmospheric Enlargement, and Size-Distance Invariance." *Perception* 29 (2000): 863–71.

Ross, Helen E., and George M. Ross, "Did Ptolemy Understand the Moon Illusion?" *Perception* 5 (1976): 377–85.

Rossi, Andreola Francesca. "*Ab urbe condita*: Roman History on the Shield of Aeneas." Pages 145–56 in *Citizens of Discord: Rome and Its Civil Wars*. Edited by Brian W. Breed and Cynthia Damon. Oxford: Oxford University Press, 2010.

Ruck, Carl A. P. "On the Sacred Names of Iamos and Ion: Ethnobotanical Referents in the Hero's Parentage." *CJ* 71 (1976): 235–52.

Ruehl, Martin A. "German Horror Stories: Teutomania and the Ghosts Of Tacitus." *Arion* 22 (2014): 129–90.

Sale, William Merritt. "Homeric Olympus and Its Formulae." *AJPh* 105 (1984): 1–28.

Salomon, Frank, and George L. Urioste. *The Huarochiri Manuscript: A Testament of Ancient and Colonial Andean Religion*. Austin: University of Texas Press, 1991.

Schaefer, Wilhelm. "Entwicklung der Ansichten des Alterthums über Gestalt und Grösse der Erde." Pages 1–26 in *Programm des Gymnasiums mit Realklassen zu Insterberg*. Insterberg: Carl Wilhelm, 1868.

Schele, Linda, and David Freidel. *A Forest of Kings: The Untold Story of the Ancient Maya*. New York: Quill/Morrow, 1990.

Schele, Linda, and Mary Ellen Miller. *The Blood of Kings: Dynasty and Ritual in Maya Art*. New York: Braziller, 1986.

Schibli, Hermann S. *Pherekydes of Syros*. Oxford: Clarendon, 1990.

Schmidt, Johanna. "Olympos." *RE* 18 (1939): 272–310.

Schulten, Adolf. *Iberische Landeskunde*. Vol. 1. Strasbourg: Heitz, 1955.

Schwabl, Hans. "Weltschöpfung." *RE*, 2nd ser. 9 (1962): 1433–1589.

Schwartz, Glenn M. "Pastoral Nomadism in Ancient Western Asia." Pages 249–58 in vol. 1 of *Civilizations of the Ancient Near East*. Edited by Jack M. Sasson. New York: Scribner, 1995.

Shapiro, H. Alan. "Jason's Cloak." *TAPhA* 110 (1980): 263–86.

Shields, Janet. "Sunbeams and Moonshine." *Optics and Photonics News* 5, no. 7 (1994): 57, 59.

Shumate, Nancy. "Postcolonial Approaches to Tacitus." Pages 476–503 in *A Companion to Tacitus*. Edited by Victoria Emma Pagán. Chichester: Wiley-Blackwell, 2012.

Sider, David. *The Fragments of Anaxagoras*. 2nd ed. Sankt Augustin: Academia Verlag, 2005.

Smith, Michael E. "The Archaeology of Ancient State Economies." *Annual Review of Anthropology* 33 (2004): 73–102.

Smith, Robert. *A compleat system of opticks*. Cambridge: Crownfield, 1738. Reprinted Bristol: Thoemmes Continuum, 2004.

Sparkes, Brian A. "The Greek Kitchen." *JHS* 82 (1962): 121–137.

Sparkes, Brian A., and Lucy Talcott. *Black and Plain Pottery of the 6th, 5th and 4th Centuries B.C.* Athenian Agora 12. Princeton: Princeton University Press, 1970.

———. *Pots and Pans of Classical Athens*. Princeton: Princeton University Press, 1961.

Speal, C. Scott. "The Evolution of Ancient Maya Exchange Systems: An Etymological Study of Economic Vocabulary in the Mayan Language Family." *Ancient Mesoamerica* 25 (2014): 69–113.

Stahl, William H. *Roman Science: Origins, Development, and Influence to the Later Middle Ages*. Madison: University of Wisconsin Press, 1962.

Steinmetz, Peter. "Tacitus und die Kugelgestalt der Erde." *Philologus* 111 (1967): 233–41.

Sweeney, Leo. *Infinity in the Presocratics: A Bibliographical and Philosophical Study*. Dordrecht: Springer, 1972.

Syme, Ronald. *Tacitus*. Oxford: Oxford University Press, 1958.

Tacitus. *Germania*. Edited by J. B. Rives. Oxford: Clarendon, 1999.

Talbert, Richard J. A. "Peutinger Map." *EANS* 640.

———. *Rome's World: The Peutinger Map Reconsidered*. Cambridge: Cambridge University Press, 2010.

Tan, Zoë. "Subversive Geography in Tacitus' *Germania*." *JRS* 104 (2014): 181–204.

Tarn, W. W. "Two Notes on Ptolemaic History." *JHS* 53 (1933): 57–68.

Taube, Karl A. "A Prehispanic Maya Katun Wheel." *Journal of Anthropological Research* 44 (1988): 183–203.

Tedlock, Dennis. *Popul Vuh: The Definitive Edition of the Mayan Book of the Dawn of Life*. New York: Touchstone, 1985.

Teeter, Emily. "Maat." Pages 319–21 in *Oxford Encyclopedia of Ancient Egypt*. Vol. 2. Edited by Donald B. Redford. Oxford: Oxford University Press, 2001.

Terio, Simonetta. *Der Steinbock als Herrschaftszeichen des Augustus*. Münster: Aschendorff, 2006.

Thibodeau, Philip. "Anaximander's Model and the Measures of the Sun and Moon." *JHS* 137 (2017): 92–111.

Thomson, J. Oliver. *History of Ancient Geography*. Cambridge: Cambridge University Press, 1948.

Tièche, Edouard. "Atlas als Personifikation der Weltachse." *MH* 2 (1945): 165–86.

Tierney, James J. "The Map of Agrippa." *PCA* 59 (1962): 26–27.

Timpe, Dieter. *Romano—Germanica: Gesammelte Studien Zur Germania Des Tacitus*. Leipzig: Teubner, 1995.

Tobin, Vincent Arieh. "Creation Myths." Pages 469–72 in vol. 2 of *Oxford Encyclopedia of Ancient Egypt*. Edited by D. B. Redford (Oxford: Oxford University Press, 2001.

Todd, Malcolm. *The Early Germans*. Malden, UK: Blackwell, 2004.

Usener, Hermann. *Scholia in Lucani Bellum Civile*. Leipzig: Teubner, 1889.

Vian, Francis, and Émile Delage, trans. *Apollonios of Rhodes: Argonautiques*. Paris: Les Belles Lettres, 1974.

Vlastos, Gregory. "Equality and Justice in Early Greek Cosmologies." *CP* 42 (1947): 65–76.

———. Review of *Principium Sapientiae* by F. M. Cornford. *Gnomon* 27 (1955): 65–76.

Vogt, Evon Z. *Tortillas for the Gods*. Cambridge: Harvard University Press, 1976.

Volk, Katharina. *Manilius and His Intellectual Background*. Oxford: Oxford University Press, 2009.

Wagner, Emil August. *Die Erdbeschreibung des Timosthenes von Rhodos*. Leipzig: Frankenstein & Wagner, 1884.

Wagner, Norbert. "Lateinisch-Germanisch Mannus: Zu Tacitus, Germania C.2." *Historische Sprachforschung / Historical Linguistics* 107 (1994): 143–46

Walsh, George B. "The Rhetoric of Birthright and Race in Euripides' *Ion*." *Hermes* 106 (1978): 301–15.

Webster, Graham. *The Roman Imperial Army of the First and Second Centuries A.D.* 3rd ed. Norman: University of Oklahoma Press, 1998.

West, David Alexander. "*Cernere erat*: The Shield of Aeneas." *PVS* 15 (1975–1976): 1–6.

West, M. L. "*Ab ovo*: Orpheus, Sanchuniathon, and the Origins of the Ionian World Model." *CQ* 44 (1994): 289–307.

———. *The East Face of Helicon: West Asiatic Elements in Greek Poetry and Myth.* Oxford: Clarendon, 1997.

———. *The Hesiodic Catalogue of Women: Its Nature, Structure, and Origins.* Oxford: Clarendon, 1985.

———. "Three Presocratic Cosmologies." *CQ* 13 (1963): 154–76.

Whitley, C. F. "The Pattern of Creation in Genesis, Chapter 1." *JNES* 17 (1958): 32–40.

Wiesehöfer, J. "Ein König Erschießt und Imaginiert Sein Imperium: Persische Reichsordung und Persische Reichesbilder zur Zeit Darios I (522–486 v. Chr)." Pages 31–40 in *Wahrnehmung und Erfassung geographischer Räum in der Antike.* Edited by Michael Rathmann. Mainz: von Zabern, 2007.

Wilkinson, Toby A. H. "What a King is This: Narmer and the Concept of the Ruler." *JEA* 86 (2000): 23–32.

Willcock, M. M., ed. *Victory Odes: Olympians 2, 7, 11; Nemean 4; Isthmians 3, 4, 7 / Pindar.* Cambridge: Cambridge University Press, 1995.

Willham, Mary Ella. "Mela, Pomponius." Pages 257–85 in vol. 9 of *Catalogus Translationum et Commentariorum: Mediaeval and Renaissance Latin Translations and Commentaries, Annotated Lists and Guides.* Edited by Virginia Brown. Washington, DC: Catholic University of America Press, 2011.

Williams, Robert D., trans. "The Shield of Aeneas." *Vergilius* 27 (1981): 8–11.

———. *Virgil: The Aeneid.* Basingstoke, UK: MacMillan, 1973.

Wolff, Christian. "The Design and Myth in Euripides' *Ion*." *HSCP* 69 (1965): 169–94.

Wolska-Conus, Wanda. *Topographie chrétienne.* 3 vols. Paris: Cerf, 1968–1973.

Woolf, Greg. "Cruptorix and His Kind: Talking Ethnicity on the Middle Ground." Pages 207–18 in *Ethnic Constructs in Antiquity: The Role of Power and Tradition.* Edited by Ton Derks and Nico Roymans. Amsterdam: Amsterdam University Press, 2009.

Wycherley, R. E. "Aristophanes, *Birds*, 995–1009." *CQ* 31 (1937): 22–31.

Xu Fengxian. "Astral Sciences in Ancient China." Pages 129–43 in *Oxford Handbook of Science and Medicine in the Classical World.* Edited by Paul T. Keyser with John Scarborough. New York: Oxford University Press, 2018.

Yang, Shao-yun. "'Their Lands are Peripheral and Their *qi* is Blocked Up': The Uses of Environmental Determinism in Han (206 BCE-220 CE) and Tang (618–907 BCE) Chinese Interpretations of the 'Barbarians'." Pages 390–412 in *The Routledge Handbook of Identity and the Environment in the Classical and Medieval Worlds.* Edited by Rebecca Futo Kennedy and Molly Jones-Lewis. London: Routledge, 2015.

Zacharia, Katerina. *Converging Truths: Euripides' Ion and the Athenian Quest for Self-definition.* Leiden: Brill, 2003.

Zanker, Paul. *The Power of Images in the Age of Augustus.* Translated by Alan Shapiro. Ann Arbor: University of Michigan Press, 1988.

CONTRIBUTORS

Georgia L. Irby is Professor of Classical Studies at the College of William and Mary. She studied Mathematics and Latin at the University of Georgia, Athens, and holds a PhD in Classical Philology from the University of Colorado at Boulder. She has published on cartography in the ancient world, the interstices of science and culture, Greco-Roman medicine, astrology, Greek pedagogy, and Greco-Roman reception in popular music. Her books include the co-edited *Encyclopedia of Ancient Natural Scientists: The Greek Tradition and Its Many Heirs* (2008); *Greek Science of the Hellenistic Era: A Sourcebook* (2002); and *A New Latin Primer* (with Mary C. English, 2015). She is currently working on a two-volume project on water in the ancient Mediterranean world, a commentary on Pomponius Mela, and the Classical influences on Kenneth Graham's *Wind in the Willows*.

Molly Ayn Jones-Lewis earned her BA from Swarthmore College and her PhD from the Ohio State University and now teaches at the University of Maryland, Baltimore County. Her research focuses on the impact that doctors and medical theories had on the larger culture of the Roman Empire. She has published on poison and political theory in Pliny the Elder and the role of the Libyan Psylloi in the Roman medical marketplace. She co-edited *The Routledge Companion to Identity and the Environment in the Classical and Medieval Worlds,* and her monograph *The Doctor in Roman Law and Society* is forthcoming in Routledge's Monographs in Classical Studies series.

Paul T. Keyser studied physics and classics at St. Andrew's School, Duke University, and the University of Colorado at Boulder, where he earned a Ph.D. in Physics and another in Classics. After a few years of research and teaching in Classics at the University of Alberta in Edmonton, Cornell, the Center for Hellenic Studies in Washington DC, and other places, he returned to his first love, programming. He worked as a software engineer for more than a dozen years at the IBM Watson Research Center and then at Google for six years in Chicago and Pittsburgh. Since early 2016, he has been working as a site reliability engineer. His publications include work on gravitational physics, stylometry, and ancient science and technology. He is co-inventor on a number of patents in computer science. He has co-edited the source book on Greek science, the *Encyclopedia of Ancient Natural Scientists*, and the *Oxford Handbook of Science and Medicine in the Classical World*. Current projects in ancient science include papers on case studies, experiments, and scientific devices, plus planned books on bottle-messages, fetal formation, the evolution of ancient science, and the history of alchemy.

Duane W. Roller is Professor Emeritus of Classics at the Ohio State University. He has a BA in Letters and a MA in Latin from the University of Oklahoma, and received his PhD in Classical Archaeology from Harvard University. He has performed archaeological field work in Turkey, Greece, Italy, Israel, and Jordan. He was also a three-time Fulbright scholar. His publications include *The Building Program of Herod the Great* (1998), *Cleopatra: A Biography* (2010), *Ancient Geography* (2015), *Cleopatra's Daughter and Other Royal Women of the Augustan Age* (2018), *Empire of the Black Sea: The Rise and Fall of the Mithridatic World* (2020), and translations and commentaries on the geographical works of Eratosthenes of Kyrene and Strabo of Amaseia. He is currently at work on a commentary on the geographical books of the elder Pliny.

INDEX